Polly Pullar is a field naturalist, w
grapher. As one of Scotland's f
contributes to a wide range of pul
Magazine, Scottish Field, BBC W
Her books include *A Richness of M*
the Highlands and *The Red Squirrel ...ature in the Forest* with
photographer Neil McIntyre. She is also the co-author of *Fauna
Scotica: Animals and People in Scotland.*

Described by the *Guardian* as 'arguably the most versatile and
inventive vet in the world', Romain Pizzi has operated on a vast
range of animal patients. In addition to his work at the Scottish
SPCA National Wildlife Rescue Centre, he has appeared on a
number of TV documentaries, including *Vet on the Loose* and *Big
Animal Surgery.*

A SCURRY OF SQUIRRELS

Nurturing the Wild

Polly Pullar

BIRLINN

This edition first published in 2023 by
Birlinn Limited
West Newington House
10 Newington Road
Edinburgh
EH9 1QS

www.birlinn.co.uk

First published in 2021

ISBN 978 1 78027 775 2

Typeset by Initial Typesetting Services, Edinburgh
Printed and bound by Ashford Press, Gosport

For Clare, with all my love,
and
for Neil McIntyre,
who loves squirrels as much as I do

Contents

The Squirrel

Whisky, frisky,
Hippity hop,
Up he goes
To the tree top!

Whirly, twirly,
Round and round,
Down he scampers
To the ground.

Furly, curly,
What a tail,
Tall as a feather,
Broad as a sail!

Where's his supper?
In the shell.
Snappy, cracky,
Out it fell.

Anon

Foreword

This book is a fascinating read, filled with anecdotes and wisdom from a life spent closely connected with the natural world. It is also Polly Pullar's very personal story of habitat restoration on her small Highland Perthshire farm and the benefits this can have for native wildlife species, including red squirrels.

From her deep fascination with nature as a child stuck at boarding school to her ongoing work hand-rearing orphaned squirrels, as well as treating and rehabilitating all manner of animals for release back to the wild, Polly shares her thoughts on the state of nature as a whole. The result is a personal but holistic picture that no academic tome or coffee-table book can offer. We share in her efforts to syringe feed tiny, pink hairless baby squirrels no larger than her thumb and worry with her about how they will fare back in the wild when they are released.

It seems strange to find myself writing a foreword to a book on red squirrels, having grown up in South Africa with its famous 'big five' – the lion, leopard, rhinoceros, elephant and buffalo. Squirrel Nutkin bore no resemblance to the Cape ground squirrels I saw. Boisterous khaki characters standing on hind legs and living in burrows, they were more like meerkats than the illustrated squirrels with tufted ears I saw dancing along tree boughs in children's books. Yet as Polly's experiences bear testimony, despite their charming appearance – exactly as they are illustrated in Beatrix Potter's stories – one quickly appreciates the tough lives these acrobatic creatures live. Not only must they survive the harsh Highland winter weather, there is also the constant struggle to find food – all the while carefully evading a host of hungry predators from domestic cats to large birds of prey. These are animals that are indeed wild, in every sense of the word.

Red squirrels now appear that very rare anomaly – a wild animal that no one in Britain maligns. Polly delves into their fascinating history, exploring how our confused attitudes to them have changed over time. It seems difficult for modern readers to fathom the sixpence bounty that was once paid for a squirrel's tail, and the fact that they were blamed for everything from killing game birds to destroying trees. Due to this misconception, squirrel-hunting societies accounted for the deaths of shockingly high numbers.

Her beautifully descriptive writing transports you into her world – a microcosm of the bigger picture – in which predators are every bit as important as the prey they hunt, despite our misguided human needs to meddle. There are no 'good' or' bad' animals. They are all part of the natural world, just with different roles to play.

Although their reputation has now been completely rehabilitated, it is sad to realise that more people in Britain have likely seen a rhino in a zoo than a red squirrel, despite their perennial popularity on Christmas cards. And we must not forget that red squirrels are still in perilous decline in the British Isles.

Filled with perception, compassion and empathy, *A Scurry of*

Squirrels is a poignant reminder for us to care for and nurture the wild, before we lose it forever.

Romain Pizzi
June 2021

Introduction

Until one has loved an animal, a part of one's soul remains unawakened.

Anatole France

I never thought I would see the day when I actually enjoyed washing-up. It has to be one of the most tedious of all domestic tasks. However, standing at our kitchen sink, I gaze out on our wild garden and see red squirrels bounding through the branches, or frenetically chasing one another as they come into the feeders. Friends have described this view as 'Squirrel Central'. Some days it's very hard to tear myself away, and I could be accused of not concentrating on the task in hand. You may be surprised to know that you can learn a great deal while you are scrubbing at the pots and pans.

In 2020 my small farm in Highland Perthshire came of age. It is twenty-one years since I bought it for me and my son, who was eleven then, and during that time it has been transformed from an overgrazed and barren hillside into a wildlife haven. As well as a great many free-living wild visitors, I also take in injured or orphaned wildlife received from members of the public or our local vets. My aim is to eventually return them to the wild, and some of these encounters will form an integral part of this story, though this is predominantly a book about squirrels.

I have always adored trees, and have since planted over 5,000, together with numerous hedging plants. I have also created a wild garden rampant with shrubs and flowers, specifically to entice all creatures great and small. Being on a south-facing hillside where we are usually supplied with plenty of rain has suited the trees perfectly. Even if the increasing frequency of our monsoon-style deluges is not always to my liking, for the trees it has been a bonus. The results are spectacular. Friends have often commented that surely I am planting trees I will never witness in their glorious maturity. However, the young trees are already stretching their boughs heavenwards. In two decades we have small, noteworthy woods. It's something that makes me immensely proud, even though in some areas of the farm, the exuberance of the trees is starting to obliterate the snaking view of the Tay Valley far below. As the trees have grown, and the habitat recovered from years of overgrazing, the red squirrels and numerous other animals and birds have returned.

There are probably few British native wild mammals, other than the red squirrel, that are so universally adored. Mention a long list that includes, otter, stoat, weasel, polecat, pine marten, badger, fox, beaver, seal, roe or red deer, and you are bound to face a wall of negativity. Someone somewhere will give you a tirade relating to the heinous crimes committed by whichever beast is being discussed. Foxes have been persecuted and reviled for cen- turies; badgers are blamed for almost every evil known to man; stoats, weasels and polecats are referred to as 'murdering little egg thieves' and are therefore loathed by game managers, as well as various other members of society. Pine martens fit the same category, and being partial to a chicken takeaway, are often simply viewed as 'poultry killers', instead of brilliantly adapted lithe and nimble mustelids with a penchant for mischief. Deer are ruinous to forestry, impede natural regeneration, and spoil the garden. And beavers, well these large reintroduced rodents are either worshipped or condemned, whichever side of the fence you perch. We have no cultural memory of the beaver's presence in our midst, and it is taking some a long time to accept the positive aspects of this extraordinary, keystone aquatic ecosystem engineer.

Though we are supposedly a nation of animal lovers, when it comes to wildlife we have preconceived ideas and like things to be orderly and to fit neatly into a category; they must be sanitised and convenient. We don't want to witness any behaviour that shocks us or seems distasteful. When that is not the case, then I am ashamed to say that humans are the most efficient and destructive killers on Earth. We wrongly classify our wildlife as good or bad, vicious or gentle, and are unable to accept that we need predators just as we need prey. For centuries we have mercilessly hunted, poisoned, trapped, snared and shot, and despite current legal protection for a whole host of species, illegal persecution still continues.

It's sadly true that if nature disrupts our lives then we endeavour to get rid of whatever it is that is causing disruption or not fitting in with our plans, in the mistaken belief that its removal will resolve the issue. However, nature hates a void, for every living thing would slot seamlessly into the bigger picture if we only let it. As soon as there is an imbalance then the entire ecosystem is unable to fully function. As we are witnessing every day through the negative world news that floods our inboxes, radios and televisions, we have pushed the natural world to the brink of total collapse.

Though you will always find someone somewhere who will tell you a damning story about a particular animal or bird, I have yet to find anyone who says anything negative about the red squirrel.

It is a fine example of a British mammal we would neatly slot into our 'good' category. Little wonder, for this glorious little arboreal gymnast instantly intoxicates us. It exudes charm, fun and devilment, it is playful – almost elfin – and instils awe as we watch it dancing around the woodland canopy. It has an enchanting little face, a beautiful coat of burnished red that glows gold in sunshine, and a superb bushy tail that it can use in numerous ways, including as a parasol or umbrella. Its little ears are often fringed with dramatic tufts that make it look as if the wearer has used the crimping tool on the hair dryer. In this country we love red squirrels with a passion; they are high on the list, often on the top of the list, of the species we most want to see. However, what is shocking is that this was not always the case.

As we struggle to protect and safeguard our small remaining population of red squirrels, it is hard to accept that up until 1981, when they received full legal protection under Schedules 5 and 6 of the Wildlife and Countryside Act, it was perfectly legal to cull them. And they were culled in such large numbers that we almost wiped them out altogether. In fact in some areas we did just that, and even now, due to loss of habitat, urbanisation and an ever-increasing road network, it is likely that many traditional red squirrel haunts have since become so ruined, or lost altogether, that they will probably never again be re-colonised by red squirrels.

Perception is a curious aspect to human nature. And everyone's perception varies. It's this that makes the way each of us in turn relates to the wildlife that surrounds us so complex. It's much the same with our incredible native flora – we battle endlessly against dandelions, which are not only flamboyantly beautiful, but most importantly are vital to pollinators including the species we love – bees, butterflies and moths – and replace them with expensive plants from garden centres that often are far less rich in nectar. In nature there is no such thing as good and bad: each living thing, great or small, is part of the magnificent jigsaw that is life itself. The red squirrel is a vital piece, and without it, that jigsaw is incomplete.

The red squirrel has always been up against it. For far too long our forebears believed that it was simply another pest, because of the perceived threat it posed to forestry. It needed to be obliterated at every opportunity. Less than a century ago, the squirrel was viewed with much animosity, but now we have taken a 360-degree turn, and the nation's swiftly dwindling red squirrel population is relishing a rebirth as one of the country's favourite mammals. Yet our darling red squirrel is not perhaps as angelic as it might appear, and many are surprised to learn that it too can predate, and is not averse to occasionally filching eggs or fledglings from an unguarded songbird's nest. Protein is an important part of its diet.

I have been fortunate to come to know and love the red squirrel from a very personal angle. I am constantly learning about its fascinating ways, its nature and its natural history. The more I learn, the more I realise that there is so much more to discover regarding these woodland specialists. Through working closely with squirrels, as well as a wealth of other mammals and birds, perhaps the most important thing I have learned is that time is fast running out and if we do not restore and protect valuable habitat, then this enchanting little rodent, and dozens of other species, may be lost and gone forever.

1

The Power of Squirrel Nutkin

I was always a child of nature and was extremely fortunate to spend part of my early childhood in one of the richest wildlife areas of the British Isles. There can be little doubt that I owe my parents an enormous debt of gratitude for living in some of the glorious places that they chose as home. Being steeped in nature from as far back as I can remember has helped to forge my undying, lifelong passion and my way of life as a writer and photographer in a bid to help others discover that precious connection with wildlife.

Fifty miles due west of Fort William lies the rock-hewn Ardnamurchan peninsula. This, the most westerly mainland point in the country, is a place of savage beauty, diverse in its land- and seascapes, surrounded on clear days by sweeping views to the Western Isles, and to some of Scotland's great mountain vistas, including the Skye Cuillin, Knoydart and the wind-scoured ridges of Torridon. Gale-battered Atlantic oak woodlands fringe the peninsula, providing an unsurpassed environment for a host of living things including rare invertebrates, flora, mosses and bry-ophytes, as well as fabulous fauna: red deer, golden and sea eagle, wildcat, pine marten, otter, fox and badger. Though at the end of the 1960s I grew up surrounded by an unequalled diversity of wildlife that I often encountered on a daily basis, the red squirrel was missing.

However, the red squirrel has been a part of my psyche for as long as I can remember, because I was reared on the tales of Beatrix Potter, and count these as amongst the very finest works of children's literature. They are highly entertaining for grown-up readers too. The fact that Potter wrote the stories and created the artwork is important. The marriage of words and pictures is unsurpassed. Her illustrations are entrancing, and often skilfully humorous. Though in most cases she anthropomorphises her animals, and her stories depict them behaving like us while dressed in clothes, her shrewd naturalist's eye ensures that each is perfectly portrayed, with an attention to critical detail that reveals very real facets of its wild character. Undress her subjects and what lies beneath is astonishingly accurate. Squirrel Nutkin has long been one of my favourite Beatrix Potter figures. Like most of the hand-reared squirrels I have been fortunate to spend time with, Nutkin is a little devil overflowing with mischief and playfulness.

When *The Tale of Squirrel Nutkin* was published in 1903, naturalist Potter probably had no idea that her brilliant squirrel portrayals and her clever words would be a catalyst that began to change our point of view. Some fifty years later, another red squirrel was a crucial player in the evolution of our attitudes towards red squirrels. Tufty became the figurehead for a road safety campaign aimed at small children. When I was six years old I was made a member of the Tufty Club and wore my badge with pride. I still have it. Now when I see an increasing number of pathetic little red squirrel bodies dead on our busy roads, it is heartrending to witness that the Tufty Club's wise message for children can do nothing to protect the vulnerable relatives of its cult hero.

It was not until I was nearly nine and sent away to a small

all-girls boarding school set on a wooded Perthshire hillside at Butterstone, near Dunkeld, that I began what has since proved to be one of my wildlife love affairs.

Hating being away from home and suffering from acute home-sickness and the instabilities surrounding the pending separation and subsequent divorce of my parents, to the staff I must have been the problem child. I was more intrigued by the birds in the garden than the lessons and I was permanently crying because I missed home and our family's large menagerie so much. A finicky eater, I hated the school food too: stodge, stodge and more stodge. We were not allowed to take food back to school with us but we could take birdseed and nuts to put out on various bird tables in the grounds, and I constantly asked my parents for more supplies. I sometimes got so hungry, having been repulsed by the proffered meals, that I ate the birds' food to stave off my hunger pangs instead. Nuts have remained a favourite food ever since. Compared to the freedom I had experienced in Ardnamurchan, now there was a strict new regime and there were also two 'big' little girls who were bullies, and who teased me incessantly. I simply wanted to go home. When I look back over it, I realise that Butterstone was not a bad place. Its setting is idyllic for a nature-loving child, as it is surrounded by oak woods and freshwater lochs where I quickly discovered that there was plenty to take my mind off my yearning to go home.

My inauguration at boarding school began in September. I remember that for the first few miserable days after term had begun, it rained incessantly. I kept diaries – hardly literary gems: *'It rained again. Missed mummy, Ginger* (my pony) *and the dogs. School food is horid. Bridget said she might be my frend.'* An Indian summer then followed this damp tear-soaked start, with balmy days that made the landscape glow as swathes of rosebay willowherb set loose their fluffy seed heads to float in the breeze.

The surrounding forests blended to a palette of yellow, ochre and bronze, and jays shrieked as they flew in bounding flight on their acorn-collecting forays, filling the air with the sound of tearing linen. Then one morning at break time when we were playing in the garden I saw a red squirrel swinging on the peanut feeder of our bird table. I remember it vividly because instantly it seemed so similar to Squirrel Nutkin, particularly as it then leapt onto the table and sat up to eat, holding nuts in what I saw as its perfect little furred hands. That night my diary entry read: '*Saw my first squirrel – it is so beautiful I cannot believe it. It looked Soooo like Squirrel Nutkin.*' And thus began another journey.

My homesickness began to ease and I even found myself eagerly anticipating playtimes so that I could go out into the garden and the school's extensive surroundings to look for squirrels. They were plentiful and the more nuts we put out, the more seemed to come. They were often there very early in the mornings but there was never time to go out before breakfast, and by the mid-morning break time they had usually gone. They had a routine and after an early start they returned at lunchtime and then again at teatime, but in winter as the days became shorter and shorter, their visits tended to be harder to time.

In the early mornings and late afternoons, fallow deer emerged shyly from the shadowy woods too. Though I was very familiar with our two native deer species in Ardnamurchan – the red and the roe – I had never seen fallow before. Their four colour variations intrigued me: common – rusty brown with speckles; leucistic – a pale cream colour; menil – with spots and dark brown around their tails rather than black; and melanistic – a really dark brown form. The gardener always complained that they ate his flowers. Sometimes he mumbled and grumbled because they also left little heaps of shiny black currants around the grand front door entrance. It always made me giggle. I had a pet sheep back home called Lulu and she did the same thing around the doorways of my parents' hotel in Kilchoan. Only sometimes she also went right upstairs and left currants outside guests' bedrooms. This was unappreciated.

We were allowed to build dens in the huge grounds, and the rhododendrons and an unruly stand of bamboo made perfect places for foundations for superb structures. However, I usually played by myself and made dens further out on the edge of the garden specially for watching the red squirrels and the fallow deer.

Living in such a beautiful landscape meant that nature education was a positive and prominent aspect to our time at Butterstone. We had a large nature table and were encouraged to bring things in for everyone to share and talk about. I asked cook for a jam jar with holes in its lid so I could keep a chrysalis I had found and we could watch it transform into a lovely moth. I excelled at nature education, art, English and reading, and was pretty bad at everything else because I found the teachers boring. We could also watch the birds and other animals from the windows, and even had walks into the out-of-bounds areas of the woods in spring to listen to birdsong. By May the cuckoo would arrive, and once I saw a tiny willow warbler struggling to keep pace with the gluttonous appetite of its obese fostered cuckoo chick. So though to begin with I hated being away from home, there were bonuses.

There were also regular school outings to nearby Loch of the Lowes. The Scottish Wildlife Trust purchased this nationally important wooded wetland site in 1969 and it has since become their flagship reserve and is in particular famous for its breeding ospreys. Soon after the purchase, a pair of ospreys conveniently turned up. On my initial school visit the spring after my first term, small groups of us were allowed to clamber up into a shoogly hide built on stilts to watch these magnificent fish-eating raptors.

Without proper binoculars or telescopes at that early stage of the reserve's development, and no cameras on the eyrie, it was primitive compared to the way every midge, every buzzing fly, the piping of an egg as it begins to hatch, and the diamond beads of raindrops on a sitting bird's ruffled feathers can be viewed today right around the globe by anyone with internet access. While the ospreys were fascinatingly exciting, particularly as the reserve's warden related the story of how these migrating birds that returned to Scotland to breed in spring had been persecuted to near extinction, it was the red squirrels that captivated me. At Loch of the Lowes some of them were so tame that they would come and take peanuts from our hands, providing we kept very still and quiet. When it involves wild things and wildlife-watching, oddly, though I am told I am a keen talker, I have never had a problem with keeping quiet. The squirrels in the school grounds were wisely wary. This was no surprise, given the racket made at playtime by a lot of shrilly shrieking little girls. However, I realised that if I broke away on my own from the team games we played in the garden, and instead went off to the woodland fringes and hid myself in my den, it was different. There I put out food on the old crumbling drystone wall, and the squirrels began to venture in closer and closer. I soon attracted enchanting little wood mice and bank voles too. One day a weasel popped its foxy little face from out of a mossy hole in the wall. Its body was like a piece of elastic. I began to recognise individual squirrels and could see that each was unique, that its mannerisms and the particular way that it interacted with the other squirrels was different too. Some had very blonde tails and flamboyant ear tufts, while others had lost these wonderful extra head adornments. Most of them preferred to be there alone and would not tolerate imposters on this new food supply; they clearly were under the impression it was theirs entirely. Squirrels are no different to people, and each has a unique character and its own particular behavioural traits. Some were more tolerant than others but in most cases, if another squirrel dared to try to venture in then it was swiftly chased away with a flurry of angrily waving tail and loud squirrel bad language. Incidentally, red squirrels have quite a temper. They are also surprisingly vocal, and won't hesitate to let you know when they are displeased, or particularly content. With fast tail flicking and irate chitter-chatter, an imaginative watcher can easily translate this to a squirrel's version of 'go forth and multiply', for that is exactly what they mean. Some of my little visitors quickly became bolder and bolder, and bounded up close to me, while others remained edgy. It filled me with a frisson of excitement that I still feel every time I view any wild thing from a close perspective.

The gardener had a large cat, and one morning it narrowly missed catching one of the squirrels as it was burying nuts on the lawn. Sometimes that damn cat would appear with birds poking from its gaping jaws, pathetic little feathers soggy with feline spittle. And it tortured and teased mice, patting at them with its fat paws, and sometimes throwing them into the air before it finished them off. And worse still, one wet morning I saw to my horror that it was carrying a squirrel's tail. I began to hate that bloody cat, and got into trouble for referring to it as such to one of the teachers. I was made to write pages of pointless lines that said: *I must not swear, because it is rude.* I have been worrying about the effects of domestic cats and dogs on squirrels ever since. Though squirrels are nature's finest trapeze artists, they are sometimes woefully un-streetwise when on the ground and come to grief all too often.

When my parents – separated by now – independently came to take me for sporadic day's out from school, it invariably included a visit to Loch of the Lowes. I had told them that the squirrels there would feed from our hands. They certainly enchanted my father during a visit one afternoon in late autumn when we were lucky to watch several of them having a final feed before they vanished back to their dreys for the night. One of them landed on his bald head. Afterwards we sat in a hide in the pinkly transforming dusk as smoky trails of mist spread over the loch, drinking tea from his thermos flask to the accompaniment of the cacophony of hundreds of wintering greylag geese splashing down onto the water to roost. That Christmas holiday I went to spend a few days with Dad. Following my parents' separation, he had moved temporarily to London and he took me to see a ballet, *The Tales of Beatrix Potter*. Squirrel Nutkin's performance on stage left me in no doubt that with the exception perhaps of the rare great crested grebe, a bird I had also seen at Loch of the Lowes, the red squirrel was indeed the finest dancer in the natural world.

2

Bough Lines

The red squirrel has been in the British Isles for a very long time. Fossil remains found at the end of the last Ice Age from some 12 million years ago reveal a similar mammal to the one we now know. The Latin name of the red squirrel is *Sciurus vulgaris* and the name squirrel comes from the Greek *skiouros*, meaning a shadow, and *oura*, a tail – shadow tail. Old local names are more prolific south of the Scottish border and include scorel, squerel, skuggie, squaggie, skoog, swirrel, squaggy, scrug, skarale and scuggery. These are likely to have evolved through the different pronunciations of our richly diverse regional accents all around the British Isles. Puggy has little similarity to the others but often appears in old documents. When I was at school and first encountered squirrels we used to refer to them affectionately as squigs, and the name has stuck. I still find myself using it on occasion as in, 'I am just going out to feed the squigs.'

The red squirrel's Gaelic names are *feòrag* and *toghmall*. Though the translation may be hazy and open to various interpretations, some of which refer to its woodland home, the best and most appropriate is 'the little questioner'. Given the extraordinary inquisitive nature of the squirrel, this translation fits it perfectly.

The French word for squirrel is *écureuil*, and in German it is *Eichhörnchen* or *Eichkätzchen* – literally meaning oak kitten, a name that aptly depicts an animal with an incredibly playful 'kittenish' spirit that is also often associated with oak trees. We use the term 'squirrel' as well as 'magpie' to describe someone who hoards or collects things. This is due to the squirrel's habit of caching and burying food for the winter. We 'squirrel' things away – special food items or perhaps precious items we don't want to share with others. Incidentally, the magpie has been known to steal items of shiny jewellery that have been left close to open windows, but the squirrel does not have the same attraction to such things. Use of the word 'con' for a squirrel is commonly found in dusty Victorian natural history tomes, and was employed in parts of England as well as in Scotland.

Collective nouns for animals and birds have always fascinated me. For squirrels it is a scurry, or occasionally you may find reference to a drey of squirrels too – which does not seem very imaginative, given that this is the name of their nest. However, I have often thought that a knot of squirrels would also be appropriate. Having hand-reared numerous litters of kits, when they are asleep together they become so entwined that it can be impossible to see which body part belongs to which little squirrel as they bind themselves tightly together as if in a beautiful tight knot.

Considering the long love-hate relationship that has surrounded the red squirrel, it is perhaps surprising that its association with folklore appears to be largely non-existent. Squirrels do, however, feature as heraldic symbols and their presence on a coat of arms, usually sitting and holding a nut or acorn, is thought to symbolise caution, thriftiness and conception. Depictions of squirrels as heraldic symbols were far more prevalent for families from the north of England; although the Earl of Kilmarnock – family name Boyd – had two squirrels on the family coat of arms, and squirrels also occur on the seal of Robert Boyd, Lord of Kilpoint in 1575, their presence for a Scottish family is thought to be unusual. While mammals such as the hare have always been revered and feared, viewed as totems, signs of good fortune, and conversely of impending doom, given both our turbulent relationship with the squirrel and its numerous charms, it seems odd that it has not captured people's imaginations in the same manner as other animals.

Ever since its arrival in Britain around 10,000 years ago, it would seem that this is an animal that has always been battling for survival. It has come dangerously close to extirpation on far too many occasions. Much of this is due to a constantly changing woodland environment, but disease and relentless persecution have played their part too. While species such as our largest land mammal, the red deer – also primarily a mammal of woodlands – can adapt to a life without trees, this is not the case for the red squirrel. Without healthy forests, red squirrels stand no chance.

The sylvan history of the British Isles is one of frequent deforestation. Since Neolithic times woods have been harvested, replanted, and often cleared entirely, initially to make way for early pastoral and agricultural activities. In the fifteenth and sixteenth centuries major timber-felling pushed squirrels to the edge, and during this era they vanished entirely from Ireland. In some places woods were even purposely razed by fire. Some were torched to drive out enemies during battles. Other enemies included the wolf. Before it was annihilated from our midst during the eighteenth century, tragically entire forests might also be burned in a bid to kill it. Inevitably, it was not only the wolves that came to a brutal end.

More recently, during the two world wars intensive tree felling for the war effort led to further deforestation on a gut-wrenching scale. During these dark periods of our history, many of the country's finest native woodlands were lost, together with a host of specialists, including the great grouse of the woods – the

capercaillie. Though Lord Breadalbane subsequently reintroduced capercaillie in 1837 to pinewoods on his Taymouth Estate
in Highland Perthshire, where for a time they appeared to thrive,
it once again faces extinction here in Scotland. The demise of our
ancient woodland cover has been progressive and continuous.
Sadly, that trend has not changed.

In Scotland, the fall of the extensive boreal forests was to entirely
transform the landscape. We became accustomed to seeing vast
open tracts of moorland and to view these wet deserts as the
norm. Even now we continue to see them as savagely beautiful.
This is the real Scotland. This is Scotland in the raw. But is this
a commensurate reflection of what would once have been here?
Previously, in many parts of the country, a richness of native trees
including Scots pine, oak, ash, hazel, willow, birch, juniper, rowan
and holly would have spread their lush mantle to cover glens and
valleys, painting entire hillsides with a palette of glorious hues
that altered through the seasons. These were places that provided
a vibrant mosaic of mixed habitat, supplying food and shelter for
a myriad of species, including the red squirrel.

An old saying states that a red squirrel could once travel freely
from Lockerbie to Lochinver without ever touching the ground.
All around the country there were similar sayings, including
this one from the Cotswolds – *A squirrel can hop from Swell to
Stow without resting his foot or wetting his toe*, and another from
Cheshire – *From Birkenhead to Hilbere a squirrel could go from
tree to tree*. Even if statements such as these are only partially true,
they highlight the extent of our sylvan losses all around the British
Isles. For squirrels and numerous other specialists, natural wooded
corridors, as well as extensive native woodland and hedgerows of
mixed ages, hold the key to their survival.

When I speak about woodland and the pressing need for continual planting, I am frequently met with the response, 'but there

are plenty of trees, so what are you making all the fuss about?'
And as we travel around the country we may indeed be under the
illusion that there are woods everywhere, and that therefore they
must be brimming with wildlife. We may be forgiven for thinking
that vast tracts of land lost under the cover of tightly packed dark
trees means that we have plenty of forest. But the hard fact is that
only 13 per cent of the UK's landmass is forested compared to
37 per cent of Europe. In Scotland, only 19 per cent of the country
is wooded, but worse still only approximately 4 per cent of that
consists of native woodland. It's a very low figure. Worryingly low.

The arrival of commercial forestry might have been seen initially
as a positive aspect to our woodland history, but in some cases due
to the attraction of government tax break initiatives during the
1970s and 1980s, it led to blanket planting regimes with non-
native conifers such as Sitka spruce. This was to prove detrimental
to most species. With their uniform ranks, and fast-growing habit,
all light was quickly excluded. In such conditions a forest floor,
usually so fecund and dynamic, transforms to a sterile and unin-
viting environment. Without a healthy understorey and vibrant
plant and insect communities, few animals and birds are able to
exist. If you are unaware of these problems you may think that
commercial planting schemes provide havens. The truth is that
the exact opposite is the reality. I think of red squirrels as litmus
paper indicating the health of our environment. If the habitat is
right for them, then it will be right for numerous other species too.

With such paucity of suitable habitat, current estimates as of
2020 suggest that there are only around 140,000 red squirrels
in the UK. And it's a number that continues to fall. However,
counting red squirrels is not the easiest task, and it's certainly difficult to be accurate, so this figure is indeed at best only an estimate.
It could even be far lower. Scotland remains their stronghold and
holds the bulk of 75 per cent of the population.

In England and Wales red squirrels are just about managing to cling on by their sharp little claws, and are still found in small, vulnerable populations, including those of the Isle of Wight, Anglesey, and Brownsea Island, and there have been introductions to a few other safe islands; there are reds in some woods and forests of Northumbria and Cumbria. Positively, commercial forestry is now changing and new woods that are grown specifically for timber are a diverse mixture that includes important native trees too, providing bough lines to help enrich the wooded environment.

Countrywide, it's not only the loss of valuable verdant native woodlands on which the squirrel and a host of other specialists depend that has led to the current decline. The situation is complex, and there are many other serious threats too. For centuries the red squirrel has indeed ridden a roller-coaster. Inevitably, there have also been natural population crashes, due largely to weather-related circumstances: winters of deep snow, or protracted periods of ceaseless rain, all of which make life for a fragile little squirrel extremely challenging. However, in these instances, when the population has plummeted, numbers will eventually begin to build up naturally again when climate conditions and food availability, particularly the production of suitable tree seeds, are conducive.

Life for the red squirrel then is far from certain, and it is likely that we will continue to hear both good news and bad regarding its situation. Our hopes for its secure future may be raised and then we are made aware that it is facing yet another seemingly insurmountable obstacle simply in order to maintain its numbers. Currently, population expansion in many areas is challenging, despite the dedicated work carried out by squirrel groups and conservation bodies all around the country who are constantly trying to improve the situation.

As is usually the case with most problems for wildlife, it is humans who continue to have the greatest impact on our native squirrel population. With the ceaseless industrialisation of our landscape, and the continuing sprawl of the urban environment, an ever-increasing road network and the insidious threat of the introduced North American grey squirrel that now numbers an astonishing estimated 2.5 million, the red squirrel has much to contend with. Roads account for the deaths of many thousands of animals every year, and young squirrels are particularly vulnerable. In some areas, rope bridges slung high across trees on either side of blackspots and known squirrel crossing areas can help, but fabrication and erection costs are high, and sadly often councils are not prepared to invest heavily when the population of squirrels in a given area may already be low. What price a squirrel? Often concerned locals and wildlife groups work hard to raise funds or to campaign for special squirrel signs, or rope bridges to be put in place. Even if only a handful of squirrels are saved, then surely it is worth it? Recently, lifelike model squirrels placed close to the road edges have been found to slow some drivers – but not everyone is either that observant, or that caring. In some cases these models have also been stolen.

It is vital to remain positive, though I admit often to finding it extremely hard. In Scotland, we now recognise that the beautiful contorted wind-blown 'granny' Scots pines that we see on post-cards and calendars are in fact trees that have come to the very end of their long lives. Some may be so old that they have even witnessed the presence of the wolf. If only they could tell us of the incidents that they must surely have seen. The pressures of over-grazing from too many deer and sheep mean that these primeval pines cannot naturally regenerate without our intervention. We are beginning to acknowledge the need for a rich forest network, including the need to plant mixed hedgerows to provide habitat connectivity that can help our precious wildlife to travel in safety to find food, shelter and a mate. All over the British Isles, a growing number of charities, as well as private landowners and individuals, are carrying out exemplary work to reinstate natural ecosystems.

Mention the word rewilding and many people think it is all about the return of missing species. This is not the case – rewilding is simply another term for restoration ecology. And rewilding instils hope for a better future for every living thing – including us. For far too long we have viewed ourselves as superior beings above nature. We must remember that nature doesn't need us, but we cannot survive without nature.

In Scotland, the last vestiges of ancient Caledonian forest at places including Glen Affric, Glenlyon, Glen Feshie, Glen Tanar, Rothiemurchus, Glen Moriston, Glen Strathfarrar, Abernethy, the Black Wood of Rannoch, Alladale and around Loch Maree are now recognised as being amongst Scotland's most precious assets. Work by Scotland's largest volunteering conservation charity, Trees for Life, has already restored vast areas that were literally dying on their feet and unable to naturally recolonise due to the indiscriminate mouths of sheep, and an ever-burgeoning deer population. Their remit too includes the translocation of red squirrels to new woodlands that they would be unable to reach naturally, due to the extensive treeless moorland that lies between.

As a passionate advocate of protecting the natural world and of restoring our beleaguered ecosystems, it is indeed hard at times to remain upbeat. I find myself constantly overwhelmed by the negativity that assaults us from all sides. However, positivity is the only way we are going to address the problems we have caused, and there can be no doubt that in recent years the shift in human attitudes is an exciting prospect for the future. In amongst the gloom of depressing news that bombards us, we are witnessing extraordinary changes in our way of responding to nature and the environment. For the red squirrel that has to spell hope. It is that hope we must grasp on to if we are to help not only the red squirrel but also the other creatures that share its world.

3

Cull of the Wild

It's so hard to imagine now, when we have raised the red squirrel to a pedestal, that for a long time it was seen as the ultimate woodland pest, a little fiend that damaged forestry, and another varmint that required culling. Coming from the species that invented clear-fell, and continues to cause havoc with indiscriminate tree felling and valuable scrub clearance, these are accusations that smack of hypocrisy. It was also previously thought that the red squirrel existed mainly on a diet of songbird nestlings and eggs. Claims of squirrel misdemeanours were rife. Most landowners and foresters had a bad word and a negative anecdote to relate about this gymnastic little member of the rodent family. *The Farmer's Magazine* of 1802 claimed that 'The squirrel is one of the most destructive animals which frequents our forests.' The poor squirrel certainly had a bad reputation.

In his *The History of the Squirrel in Great Britain*, published in 1881, naturalist J.A. Harvie-Brown includes numerous stories and supposed evidence of the numbers of squirrels killed. He also includes fascinating viewpoints from many of the sportsmen/countrymen of the day. While Harvie-Brown himself was one of the few who were in favour of the squirrel's presence, he was not their only friend, as his quote from the Reverend Lachlan Shaw's *History of the Province of Moray* (1775) proves:

> There are still in this province foxes, badgers and squir-rels, weasels etc. The squirrel is a pretty, sportive harmless creature: it is a kind of a wood wesel (sic) – haunts the fir tree. If you toss chips or sticks at it, it will toss pieces of the bark back again, and thus sports with you. If it is driven out of a tree, and skipping into another, finds the distance too high, it turns back to its former lodge, its bushy tail serving as a sail or wing to it.

It is a rare positive comment; Harvie-Brown states that an incensed writer in *The Field* magazine of 1878 claimed that squirrels attack firs, ringing their tops and causing rot, leading to the wind breaking them off, while Sir Dudley Marjoribanks of Guisachan, Inverness-shire wrote: 'The old people say the squirrel drove away the red-headed woodpecker from Guisachan.'

It's a bizarre claim given that in our garden, and doubtless many others, there is a healthy population of both greater spotted wood-peckers (the one Marjoribanks refers to) and squirrels, and neither takes any notice of the other. However, when it comes to nature's vandalism, the extensive joinery work carried out by the wood-pecker on bird nesting boxes *is* a constant issue, and frequently we watch as the occupants are pulled out of their maternity suites and put on the menu for the woodpecker's own brood. Other writers of the era claimed that they found squirrel dreys filled with the heads of game birds. I do not buy this and feel that it would have been most convenient to make such accusations if you wanted your peers to join forces and find reasons to dislike and vilify the squirrel.

Sadly, to this day anything that is a threat to the sacred game bird is no friend to those who run shoots and sporting businesses. This too is ironic, given that numerous species are 'culled' to protect game, which finally meets its end at the point of the same gun. We humans are hard to fathom. Other dead birds have been found in squirrel dreys. I found a record from the particularly harsh winter of 1878–79 when a group of coal tits were found in a drey. The likely explanation for this is that the starving birds went into the mossy haven to shelter, and instead quickly succumbed to the intense cold. Communal roosting is not unusual amongst small birds in sub-zero temperatures. Dozens of wrens may huddle together for warmth and also often die despite congregating close to safeguard themselves against the climate. However, those who hated squirrels probably preferred to blame them instead, even though apparently the coal tits had not been eaten and only desiccated corpses remained. Reading attitudes from the past, it seems as if anything that could be dredged up and the facts modified to turn people against the red squirrel, was – and frequently.

In defence of the squirrel, Harvie-Brown quotes a Mr Peter Scott, who wrote to him further about their misdemeanours. 'As to their taking game or eating carrion I am not sure only I have known of ugly things being found in their nests, such as a pheasant's head, rabbits' and other kinds of bones.' Harvie-Brown adds his own comments. 'This evidence appears to me in no way whatever to prove a carnivorous desire, but simply the gratification of that inherent curiosity, mentioned before. Much more likely that a weasel, or stoat, or other carnivorous animal slew the pheasant, and left the head lying, and that our little friend, happening to pass that way, or having watched the weasel or stoat at its meal, descended afterwards from its arboreal perch, and pouncing on the pheasant's head, bore it away to its drey.' I would very much like to have met Harvie-Brown as he seems a far better naturalist and observer than most of those who portrayed themselves as experts back in the day.

Sir Dudley Marjoribanks of Guisachan seems to have had a particular fascination with squirrels and Harvie-Brown also quotes him saying that, 'The squirrel here is a shabby little animal in comparison with his southern brother. I doubt whether he would weigh half as much in the scales – and he is not nearly so rich in colour.' Presumably he had seen healthier squirrels elsewhere. Records show that there were several very exacting, hard winters at that time that would surely have taken a toll on struggling squirrel populations in Inverness-shire.

Again, Harvie-Brown has an entry of note in his book regarding the presence of squirrels in Clackmannanshire where 'a squirrel was killed on the coronation day of Her Majesty – June 20, 1837 – by one of a party of some twenty bark-peelers in the Dollar district'. He goes on to relate that, 'In Clackmannan, we find that, while unknown around Alloa, when the "Old Stat. Account" was written, it had become numerous by 1841. The Rev. P. Brotherstone, who writes a somewhat able article on the natural history of the parish of Alloa, takes occasion to notice the service done by the squirrel in planting acorns, thus supplying future oak trees for the British Navy.' This last comment highlights that we have long regarded wildlife as something useful, but I suppose cynically I must add that at least the Rev. Brotherstone was writing something positive about the red squirrel. And that appears to have been most unusual during the mid-1800s.

There was still plenty of evil business at work. In the 1860s squirrels were said to have been so plentiful on Cawdor Estate near Inverness that it was necessary to cull them. There is a shocking record from 1867 of 1,164 killed in that year alone. On some estates, a bounty of 6d (six old pence) was paid per tail, giving keepers and farmers incentive to continue the war against the innocent animal. Some keepers are reputed to have caught squirrels, cut their tails off and set them free in the belief they would grow another. If this were the case, it would enable them to receive payment twice. There's always so much to learn about every species, but I find it particularly surprising that they believed a new tail would grow back. The only species in this country that can do that are the lizard and the slowworm – a legless lizard – though the regenerated part of the tail will never be quite the same. You will find similar miserable stories in countless books written by the sportsmen/naturalists of the time. Most, if not all, were obsessed by killing and therefore it seems a paradox that they styled themselves as 'naturalists'.

According to Harvie-Brown, 'The numbers of squirrels killed depend a good deal upon the qualifications of the men employed, and on the price paid per tail.' He adds, 'Nor can we implicitly trust such records, because as Capt J. Dunbar Brander of Pitgaveny said, "on the estate of Cawdor many thousands of tails were paid for, supposed to have been killed in the district. One day the factor saw a bunch of squirrels' tails arrive at the station addressed to one of the keepers; a day or two afterwards they were presented to be paid for". Nefarious business indeed!

Many landowners saw the chance to cash in their red squirrels in other ways too. They trapped them on various estates around the country, and then packed them up in unsuitable transport boxes and sent them by train to markets in the south. Here the unfortunate animals were put in cages to be kept as pets – a curiosity to grace the drawing room of a grand house or castle – a talking point. Most were terrified witless by the horrible ordeal. Squirrels are exceptionally susceptible to stress and disease. If this did not kill them first, then they frequently bit their captors. Teeth that are designed to crack nuts can sink deep into flesh and bone, like a knife through soft butter. Those that were unlucky enough to be so wounded swiftly became less enamoured of their new purchase and saw it instead as a fiendish red peril. This was likely to seal the fate of the blameless squirrel. It is dubious whether the idea of making money from squirrel sales proved successful for those who were cruel enough to take them from their woodland home in the first place.

Like the stoat and pine marten, the red squirrel was also highly sought for its soft fur. In Finland squirrel pelts had a high value and were used as currency in a barter system for goods prior to the introduction of currency. 'Oravannakka' – squirrel pelt – is a word occasionally still used there when discussing money. Folklorists have suggested that Cinderella's slippers were in fact made of squirrel fur and not glass. The French word for fur is *vair*. There are theories that this evolved instead to *verre*, the French word for glass.

Again according to Harvie-Brown, in 1836 John Colquhoun rented shooting at Kinnaird in Perthshire, where squirrels were particularly abundant. Fifteen years later he saw a large table at a local market with over 200 squirrel skins on it, shot on the Taymouth Castle Estate. 'They began to kill them down at Taymouth between 1848 and 1849. I was at a bazaar in Perth, and the Marchioness of Breadalbane was there and had a stall. She had about 500 skins for sale.'

Identification appeared to have been a problem, with much confusion between squirrels, stoats and weasels. In 1810 a Stirlingshire woman captured a weasel and claimed it was a strange beast. She was convinced it was a squirrel. She wanted it because she had heard that they made good pets. However, she too was severely bitten and told her friends afterwards that she had caught a tartar. In 1812 Harvie-Brown says a squirrel shot by a Mr Alexander McLean of Lochend Cottage, Almondbank, Perth, attracted numerous curious individuals who travelled a considerable distance to come and see it. Such tales would indicate that squirrels were very uncommon in Perthshire, one of their current strongholds, at this time, yet soon after they were culling hundreds. Could the population really rise and fall with such alacrity? There seems to be a discrepancy somewhere along the line. There were also reports of single squirrels seen climbing chimneys and sometimes running far out into the open on treeless moors. One gamekeeper described an incident where the squirrel in question climbed all over him. He thought the reason for this was because there was not a tree in sight. Its relative tameness seems odd even if there were no safe havens, but I have personally witnessed on a few occasions how, when in dire straits, squirrels (particularly kits that are newly emerged from their dreys), may become so desperate with hunger that they temporarily lose their fear. The Rev. W. Gordon of Braemar told of another incident relating to the squirrel's peculiar habits. He said that it was an animal forever getting into scrapes and gave the case of Mr Robert Grant, who had captured a squirrel to keep in a cage. 'On its escape immediately after, it suddenly climbed up the inside of the legs of its captor's trousers which having been "made down" were unusually wide.' Wide trousers and squirrels are not a good combination, particularly if you are a man. Don't let's forget that the squirrel has a fine set of nutcrackers.

Far earlier, during the late 1700s, several enlightened Scottish landowners, including the Duchess of Buccleuch and the Duke of Atholl, brought in red squirrels from Scandinavia. Later, others followed the trend, and Lady Lovat brought them into Argyllshire during the 1800s. So it is obvious that squirrels fiercely divided opinions. While some landowners went to great pains to bring them into their woodlands, others paid their keepers a bounty to kill them. Even during the early 1900s squirrel hunts continued. If records are to be believed, between 1891 and 1903

Atholl Estates, which had reintroduced them earlier, accounted for 3,069 dead squirrels. Squirrel hunting was one of numerous similar unpleasant pursuits with associated parties taking place all around the country. These hunts were viewed as special occasions to mark Saint's Days as well as Christmas Day and Boxing Day. Large groups gathered for these destructive woodland forays as, armed with long poles, men and boys beat any dreys they found, sending the petrified squirrels into the open often to be felled with a battery of stones from well-aimed catapults, as well as guns. A particular tree with a high drey might be cut down to persuade the squirrel to leave its refuge. In a report from the *Glasgow Herald* in 1904, the Ross-shire Squirrel Club killed nearly 4,000 in that year and accounted for almost 5,000 in the one previously. Other records claim that the Highland Squirrel Club accounted for over 82,000 red squirrels between 1903 and 1929 in a bid to minimise the damage they caused to trees. Even though attitudes were gradually changing, the hapless red squirrel was still viewed as an enemy.

4

The Arboreal Athlete

Surely no one could deny that when it comes to beauty and allure the red squirrel has it all. At a glance it is both adorable and cuddly. It is one of the most popular wildlife subjects seen on cards and calendars, and you will also find it posing sweetly as a soft toy bedfellow, complete with plush nylon parasite-free fur, luxuriant eyelashes and a beguiling smile. It will be found in the form of an astonishing range of ornaments, mugs and coasters, and when it comes to cards and calendars, especially at Christmas, the squirrel is increasingly beginning to knock the robin off the top spot. Yes, the red squirrel indeed attracts a cornucopia of fripperies for our delight – some worse than others.

We love squirrels and we do seem to view them as 'cute'. This is a word I have always disliked intensely for describing any wildlife, but I am in a minority, as it's one of the most frequently employed. You will hear the red squirrel also referred to as 'iconic', another vastly overused term, and as far as I know, the squirrel has seldom, if ever, been used as a religious symbol – the true meaning of icon. Increasingly often it is also referred to as charismatic – but I find this altogether too human. Squirrels are all these things, it's just that I feel from a descriptive point of view we could do better, and be far more accurate. There's a great deal more to a squirrel than you perhaps realise: it's not simply a charismatic, iconic cutie-pie.

If you study nature's finest mammalian athlete when it's in full moult, when it can look as if its dapper little red jacket – usually the colour of burnished conkers – has been attacked by moths, then you will briefly see that this is the time when it may momentarily resemble a tree rat, as portrayed by our Victorian ancestors. But only momentarily!

Rats generally cause us to recoil in disgust. The old adage 'you are never more than six feet away from a rat' may not be strictly accurate, but there are plenty of these loathed rodents living in close proximity to most of us. Though we find this horrible, ironically the reasons for their high number include the messy manner in which we conduct our consumerist lives. Yet even though rats are so common, most of us know very little about them, yet we cling on to a smattering of ill-conceived bad press: they are filthy, disease-ridden and disgusting. However, most of this is inaccurate. If we ever for a moment stopped squealing in terror while racing for the nearest chair and instead lingered long enough to study a rat's conformation, we would see that though squirrels may appear similar when they have lost their ear tufts and are suffering from seasonal baldness, they are very different. The two don't have much in common other than perhaps the long tail (but the rat's is bald and scaly) and a cheeky rodent face with prominent front teeth. Compared to a rat, a squirrel's body is long and lithe, and is all held together with an exceptionally flexible, strong spine. Its forelegs are relatively short compared to its incredible elongated, powerful hind legs, with their specially adapted double-jointed ankles. These allow the squirrel to power its Olympian feats of long jump. Even when newly born, kits have surprisingly sharp little claws. Sometimes they accidentally scratch one another, and often scratch their mother's mammary glands while suckling. Both claws and ankles are modifications to enable a squirrel to climb and to grip on to almost anything.

An elderly forester friend who knew a thing or two about nature once told me, 'A squirrel could easily climb a stairway to heaven, if only there was one.' It's an idea that paints a particularly beautiful picture and accurately portrays an ability to skip out onto the tips of the very highest and thinnest branches, using that flexible tail like a trapeze artist's balancing pole. Yet sometimes even a squirrel may come to grief. On occasion I have received unfortunate animals with broken limbs, or perhaps a smashed hind foot or front paw. Often they have been found in woodland far from roads, so it seems likely these are climbing injuries sustained where aerial leaps have been misjudged. Even if it's possible to mend these breaks, which it usually isn't, it is dubious whether it is sensible for the afflicted to return to the wild. This means that the fall will inevitably cost the animal its life, and certainly its freedom.

When it comes to big feet, it's one of the most noticeable physical characteristics of even the youngest squirrel kits. Their feet always appear several sizes too large and remind me of small children wearing capacious welly boots. Squirrels never seem to grow into their feet but it is these, with their long toes and that elasticated body, that aid their agility, allowing them to hold on and dangle upside-down, hanging on literally by a toehold. The digits of their front paws are elongated too, and are relatively many times longer when compared to the fingers of a human hand – all the better to grip on to bark, branch and trunk.

Perfectly suited to their sylvan lifestyle a red tends to do most of its eating aloft, high in the swaying woodland canopy, while a grey squirrel tends to eat more on the ground. With a very fast metabolism, it must eat regularly in order not only to survive, but to keep up its energy levels for its frenetic lifestyle. In between meals – and some days a squirrel seems to feed incessantly – a squirrel needs to sleep. Squirrels seldom, if ever, suffer from insomnia, but they are particularly early risers, and as the days grow longer they may be abroad as early as 4 a.m., feeding to the accompaniment of the dawn chorus. They will then have regular siestas throughout the day, and usually go to sleep well before dark. In our garden the resident squirrels feed early and disappear during the morning, only to return again around lunchtime. Then there is a flurry of feeding activity around late teatime too. The majority of each day is spent searching for food, and in between times a squirrel will be off to snooze quietly in one of its dreys.

Though the red squirrel is a specialist that thrives in pine forests, it has a varied diet that is governed largely by the seasonal availability of nature's larder. The small seeds found inside cones such as Scots pine and Norway spruce are a favoured part of the diet, and many other species of tree seeds are also consumed. Its swift metabolism means that starvation may only be a few missed meals away. Unlike some other animals, a squirrel does not carry excess fat reserves. For a gymnast whose life depends on aerial agility and precision accuracy, excess weight would lead not only to loss of energy, but also restricted mobility.

Spring is a good time for a squirrel as buds and shoots are emerging and trees are abundant with blossom and sap to add to delicacies such as hazel catkins and lichen. It may come as a surprise to learn that high summer is not a time of abundance. In fact, it can be very hard to find enough food. Now buds and shoots are gone and instead trees are fully clothed in heavy foliage. Nuts, seeds, fruits and berries are still to ripen. This is when garden feeders can provide an important source of sustenance, particularly for young squirrels that have newly emerged from their natal dreys. In summer our squirrels frequently raid the vegetable patch too – peas and beans are filched with visible glee, and sometimes little paws uproot carrots, much to the pleasure of the destructive carrot fly that can then get in to the roots to cause mayhem and subsequent ruination to my precious crop. By early autumn woods and wild gardens provide a storehouse of squirrel exotica: hazelnuts, acorns, berries, fruit, fungi and seeds are all relished. And in our garden I watch as little red flashes skip back and forth collecting nuts to cache ready to sustain them through the winter. Often another squirrel gets there first and is swift to pilfer the stash, and there may be a tussle amid impassioned chattering. Sometimes I may find some of their hidden stores in my flowerbeds and flowerpots – and often the planter removes bulbs in the process. There's no denying the culprit, as neat tooth marks leave evidence – the bulb has been tried and tested but deemed unpalatable, and is then discarded with disdain. There are beneficiaries lying in wait, and now wood mice and bank voles take advantage as new tooth marks are added to my battered bulbs. When it comes to collecting and burying acorns, both squirrels and the shyest and loveliest member of the corvid family – the jay – bury copious amounts but lose track of their burial sites. In this way, providing the area is free from the efficient mouths of deer and sheep, entire new oak forests may take root. Nature's planters have an important role to play.

Bark is expertly stripped to reveal sugary sap beneath, and later in the season an assortment of fungi is collected and stored, and often left out to dry, wedged into the fissured bark of a tree trunk, or taken to the drey to be consumed during hard times ahead.

As omnivorous rodents, diet is surprisingly catholic and other items on a squirrel's menu may come as a surprise. During the summer, when food may be scarce, it is not unusual for a squirrel to consume both the eggs and nestlings found in an unguarded bower that it discovers on its travels through the high tops. It may not always fit with our vision of a 'sweet, charming' little woodland sprite. It's important to recognise that the small amount of bird-based snacks consumed by a red squirrel do nothing to dent the avian population. In the same way as a goshawk, buzzard, carrion crow, fox, or pine marten may kill and consume a squirrel, so too a squirrel may occasionally add the delicacy of a songster to its diet. Squirrels have also been seen feeding on carrion. When I put out dead day-old chicks for our resident tawny owls, I sometimes see the squirrels sampling them, though it's an infrequent occurrence. I recently had an interesting, and perhaps equally surprising encounter in our garden with a blackbird. We have a nursery roost of pipistrelle bats in the old bothy close to the house. In the midge-ridden gloaming of high summer they emerge mysteriously, spilling forth into the gathering darkness. During the day I hear their high-pitched sounds coming from beneath the slates, and smell the strong, distinctive aroma of bat guano. I watched as a cock black-bird pulled a tiny baby bat from beneath the slates, and it rolled down the roof onto the ground. Within seconds the blackbird was on to it, and I sat in silence, fascinated as it devoured the unfortunate bat. The point here is that everything will eat something – it's an integral part of nature's vibrant tapestry. Though the black-birds scold aggressively when there is a stoat, weasel, pine marten, cat, sparrowhawk, woodpecker or any member of the crow family around, they never bother to give voice when squirrels are even very close to their nests, indicating they do not feel particularly threatened by their presence. Yet as we have seen, the enchanting elf-like red squirrel is a Jekyll and Hyde. Life in the wild is never easy, and a squirrel, like every other creature, must adapt in order to survive.

When I am hand-rearing squirrels it's vital to ensure they have plenty of vitamins and minerals, and I feed them a variety of fruit, vegetables, nuts and seeds; as well as many of the natural seasonal foods they would find in the wild, I am always interested to see what other foods they favour. Melon, grapes and fresh sweetcorn, sprouts, celery, cucumber and blueberries are just some of the more unusual favourites. When it comes to nuts, they go mad with visible glee for pistachios and if there are a few in their mix, will instantly ferret through the dish with little paws until they find one. These nuts can never have formed a part of their natural diet. They do love peanuts too, but it's important not to let them become too dependent on them and to give them choice, as peanuts alone would not provide a balanced diet and are lacking in important minerals.

Like all mammals, squirrels require a source of fresh water. They may also drink from water-filled leaves, as well as boles and cavities in a trunk high in the wood. A burn runs through our garden and on hot days I have watched squirrels drinking at the water's edge. Though squirrels are proficient swimmers, it seems unlikely that they enjoy it. On the shores of a freshwater loch near Dunkeld, I once saw a small head protruding from the water surface far out into the loch and thought at first it was a stoat or weasel. Whatever it was made steady progress towards me across the still, glassy surface on a moody July afternoon. When it reached the bank and emerged from the water, I was surprised to see it was a squirrel. At that moment no better description fitted it than that of the proverbial drowned rat. A sunlit spray of beaded pearls sent droplets into the air as the diminutive squirrel shook itself before bounding past me, oblivious. In those fleeting seconds I saw the true length of its little form, the fact that it carried no excess baggage, not a gram of spare fat. I could see no reason for its swim. Vulnerable to the last it vanished into a straggle of Scots pines, leaving me in a pensive mood. Nature is full of surprises. But why shouldn't a squirrel swim? However, I imagine it had been frightened and therefore forced into taking such action. I am convinced that as this is such a rare sight a squirrel does not take the plunge by choice.

When the trees are in full leaf, it's far harder to watch squirrels in a woodland environment. I love to go deep into the woods, to places where the squirrels discard their partially eaten cones and hazelnuts, and find a stump to sit on and start my vigil. In high summer the woods are drowsy with humidity. Some days the stillness is audible. Birds once resplendent in full breeding finery are now hiding away as if embarrassed by their heavy moult. At this time they are vulnerable, following the rigours of the breeding season. Their joyous melodies are but a whisper, though the melancholy song of the robin and the sweet sonata of the wren still waft into dark green glades. I may have to wait for a while, but soon the first sounds of chewing filter down from above. Squirrels are noisy eaters, and sometimes when they know they are being watched, they may swish their tails and chakk in irritation. While I may be able to hear them, finding where they are perched can take a few moments. Then there it is, that neat rounded shape with tail looped over its back, sitting in a sunbeam while casually chucking unwanted bits of cone in my direction.

Squirrels are equipped with the best teeth in the business. They grow continuously and wear down, providing their jaws are perfectly aligned. These are teeth, precious tools, with the ability to work tirelessly as nutcrackers. It takes great skill as well as dexterity to extricate the small, important seeds from tough spruce and pinecones. A cone is first examined in the forepaws, turned and viewed from all angles for a general assessment (like us, squirrels tend to be left- or right-handed). Once deemed edible, it won't take long before the squirrel's work is done and it quickly moves on to the next goal. There's nothing laid-back about a red squirrel – everything it does involves agility and speed.

In a healthy woodland environment there are numerous symbiotic relationships between animal and plant communities that share the habitat. Deer shed their antlers every year: roe bucks lose theirs between October and December and grow new ones over the winter months, while red deer stags shed theirs in March and April. The larger and more impressive the animal, the earlier it will cast its antlers – that is to say that those with the biggest antlers usually lose them first. Antlers are never wasted, for once shed they provide a valuable source of calcium, as deer and other animals chew on them. Particularly for the pregnant hinds and does, they provide a rich boost of minerals at the end of winter when they may be at low ebb, and therefore greatly help with the healthy development of their growing foetuses. Antlers are also valuable for squirrels for the same reason, and they are magnetically attracted to them. Tooth marks from both deer and squirrels are often clearly evident on antlers that I find on the forest floor above our farm. We have been putting up antlers in our garden for our visiting squirrels for many years, usually fixing them in an upright position on top of our aviary roofs so that they cannot be dragged away. These natural sculptures provide us with a surprising amount of entertainment as they are thoroughly investigated and used as climbing frames, or as a high and precarious place from which to tick off any other squirrel that dares to come close. Some of the youngest squirrels also enjoy honing their climbing skills while chasing one another around like whirling dervishes playing tig, having first tested out their new teeth. Dried cuttle fish will also provide a mineral boost for squirrels and are put out by some garden owners. We have tried both, but there is no doubt that when there is a choice, antler is the favoured option.

*

When Beatrix Potter portrayed Squirrel Nutkin and the other squirrels sailing across to Owl Island to visit the tawny owl, Old Brown, her picture showed them using little rafts made of twigs, and employing their luxuriant tails as sails. It's not so very far from the truth. For a squirrel, a tail is of vital importance and it can be used for numerous different purposes.

Not only does the tail help the squirrel with balance, but it is also used like a rudder to add direction and give momentum to those enormous leaps of faith. In our wet climate, it serves perfectly as an umbrella too, while on hot days it becomes a parasol. It is always amusing on a day of heavy rain to watch a squirrel at the peanut feeder battling against the elements with its tail over its head. A good thick tail can also help to keep a squirrel warm in its drey, where it serves as a superb blanket. Tails are sometimes employed as windbreaks when their owners are sitting high in a tree on a blustery day.

Tails reveal a great deal about a squirrel's mood too, as they are held in different positions depending on how relaxed or nervous an individual feels. One of the most important ways in which a squirrel uses its tail is as a flag, to signal to other squirrels exactly how it feels by waving in either greeting or warning – usually the latter. When a squirrel encounters something that makes it unsure or frightened, particularly a dog, cat or perhaps a person, before it flees in a hurry there will be much tail-waving, accompanied by questioning chatter.

Sometimes when I am filling up an empty feeder, one of our little squirrels will bravely sit quite close, emitting a range of sounds of displeasure, while waving its tail almost as if it is annoyed that I let the feeder run dry of nuts. While walking in woods with my three collies, far from the resident garden squirrels that have grown accustomed to me, I often stop to watch a squirrel, and it too will wave its tail crossly, warning us to keep away. For a squirrel, the tail is a vital means of expression and communication, and reveals much about its humour and disposition.

At a glance when we are watching the individual squirrels that visit the garden, their tails also provide undisputed confirmation of identity. There is much colour variation, particularly during the summer months, when a tail can bleach and turn quite blonde. In 1905, the painter and naturalist John Everett Millais noted, 'From autumn to March their tails bleach, becoming cream or even white.' I have found, however, that this bleaching also tends to take place during the sunnier months. And some squirrels have very dark tails, or even racoon-like banding on them.

A few years ago we had regular visits from a female that only had half a tail. Surprisingly, she survived perfectly well and continued to visit for approximately three years. We often wondered how she had lost it. At the time we had a particular cat in the local area that was lethal for squirrels, or perhaps she had been grabbed by a large bird of prey. Being caught in a fence or a trap was another possibility. Whatever had happened, she had been lucky to escape. Though she came close to the kitchen window to feed, there was never any sign of a wound; this was clearly an old injury and was well healed by the time we first encountered her. Thankfully, it did not seem to impede her agility, though that surprised me.

When you glance at a squirrel you may be under the illusion that they are all the same colour, but this is not the case. Look a little closer and every one is an individual. In summer there is little uniformity to their glossy new jackets. Some are a deep, dark shade of red, almost auburn, while others are varied tones of ginger, rust, amber, gold or bronze. Then occasionally there are those that look as if they have been to the hairdresser to have blonde highlights added to their pelts. Occasionally someone may send an image of a pure blonde, or even an albino, that is visiting their garden. Leucistic red squirrels are not unusual. Leucism is a term used to describe a wide variety of conditions that result in the partial loss of pigmentation in an animal or bird. It causes white, pale, or patchy coloration of the skin, hair, feathers, scales or cuticles, but not the eyes as in albinos. However, because these unusual animals

stand out, it would seem that they are more vulnerable to predation. As mysteriously as they appear, their disappearance is equally
mystifying. Being so conspicuous is not advantageous.

Red squirrels in other parts of Europe also have a wide variation
in hues. All the squirrels in the British Isles are of mixed genetic
stock following frequent reintroductions in the distant past.
Genetically our red squirrels are not distinct from those found in
the rest of Europe. However, some say that during the time that
different red squirrels were introduced to this country from other
parts of Europe, colour variation became more prevalent, and suggest that this is the reason for such a wide range of shades in our
red squirrel population today. As a wildlife rehabilitator, I receive
many injured tawny owls that also have dramatic colour variations
ranging from the richly red-rufous-phase to grey-phase birds.
Buzzards too can be very varied. I am inclined to think that it is
simply nature, and like humans all red squirrels are individuals.
After all, nature does not regiment anything.

Squirrels have spectacular long whiskers and some have far
longer and more impressive ear tufts than others. A man who
claimed to be a squirrel expert once told me that you can sex a
squirrel by its ear tufts – 'Males have them and females don't,' he
announced with authority. I am not sure where he got this idea
from – it's a fallacy – but I did not argue the point. All squirrels
may have tufts and there does not seem to be any rule as to why
some have more luxuriant ones than others, and certainly does
not ever relate to the different sexes. Equally, some squirrels retain
their tufts in the summer, while others lose them. Usually squirrel
kits have tufts throughout their first summer, but one litter that
I hand-reared had no tufts at all, while another litter grew very
impressive ones, and both litters were reared at exactly the same
time of year – early summer. In summary, like colouration, there is
no uniformity and sometimes it's not possible to find the answer,

which is another reason why close study of anything to do with nature is so intriguing. Tufts add to a squirrel's charm. My friend the wildlife photographer Neil McIntyre, who has a world-class collection of red squirrel images, has some incredible pictures of squirrels taken in the woods around his home near Aviemore in the Cairngorms. They have fabulously long ear tufts and are amongst the most attractive squirrels I have ever seen, especially those that appear coiffed. Like tails, tufts can vary in colour and may be paler or darker than the rest of the wearer. Tufts are a charming addition to the perfection that is the red squirrel.

5

A Drey in the Life of a Squirrel

Great fleas have little fleas upon their backs to bite 'em,
And little fleas have lesser fleas, and so ad infinitum.
And the great fleas themselves, in turn, have greater fleas
   to go on
While these again have greater still, and greater still, and
   so on.

Augustus De Morgan

For a squirrel, a drey, the correct term for its nest and home base, is a very important place. Squirrels make numerous dreys throughout their short lives; most squirrels average only three to seven years in the wild, but can live much longer in captivity – sometimes up to ten years, or even a little more.

Dreys vary considerably and are found in a wide variety of tree species. A fork is often a favoured place and makes an excellent framework for building a safe nest. Hollow trees and old wood-pecker nest holes are also chosen. Sometimes dreys may be built around an abandoned crow's nest. The higher the site, the safer the squirrel will feel. Though dreys are usually sited away from prevailing winds, sometimes in lofty locations, torrential down-pours and high winds can prove perilous, and many dreys are blown down. Usually the occupants escape. However, when a drey with kits blows down, it can be more serious. Usually, the mother will quickly move her young to a new one close by. In these instances, if you are lucky to be in the right place at the right time, you may even witness her racing along carrying a kit by the scruff of its neck, its little body swinging back and forth as she hurries to relocate her litter, and gallops back again for each one.

Summer dreys are more flimsy and less well built than those made for the winter. In the colder months dreys tend to be structurally solid and luxuriantly lined with sheep's wool, hair, moss, lichen, or feathers, and often a mixture of all these materials. Basically, what-ever the squirrel can find will be employed to make the drey cosy. For the basic construction, twigs, leaves and lichen are used and then folded around branches in a similar way to nest fabrication by birds. However, squirrels never seem to be so creatively ornate. One drey that I found was thickly lined with our dogs' hair, while another, discovered in hazel woods close to my home, was not only still bouncing with fleas, indicating that it had recently been occupied, but it was also lined with horse hair gathered from the adjacent field. I have found dreys richly lined with soft pigeon breast feathers perhaps plucked from a carcase, or collected from a sparrowhawk's plucking post – the spot where it has taken its prey and consumed it, leaving heaps of feathers behind. Squirrels are innovative. And they do like their comfort. In the spring of 2020, when we had a protracted spell of the finest weather in memory, I sat out on a bank of wildflowers in our fields and watched a little female laboriously gathering moss from the field margins. During the hot, dry spell, badgers had been searching for invertebrates and had been digging and pulling the moss from large rocks on the bank. It was the perfect material for a new maternity drey and was an excellent example of how animals work in harmony with one another and take advantage of the situation.

Where there are no mature trees, squirrels will build their dreys in low bushes and shrubs. The end result may dangle conspicuously, like a weaverbird's nest in an African thorn tree, yet these structures are usually surprisingly robust too. The greatest problem is that dreys in a low position are not always safe from other woodland residents, and are vulnerable to predation, particularly from members of the Mustelidae family, domestic cats, goshawks, buzzards and corvids. During the winter months, when in a deciduous woodland trees are skeletal and bare, it is far easier to locate dreys; however in a coniferous forest, with its palette of dense dark evergreens, dreys built in a Scots pine are usually sited high and are almost impossible to see.

Though when they are out and about squirrels don't seem to enjoy one another's company, and frequently chase one another away from a food source, a few may often sleep together in a drey, particularly when the temperature plummets. It makes sense to share body warmth, and as I mentioned earlier, that versatile tail works well as an excellent blanket to wrap around yourself or your sleeping companion. Squirrels don't hibernate, though many people are under the impression that they do, but during periods of excessively bad weather they have a tendency to sleep for longer periods, simply to reserve energy levels.

Mention hedgehogs and often one of the first thoughts is that they have a reputation for being fleabags, yet this is something you seldom hear anyone mention in relation to squirrels. However, they too have fleas – copious amounts. It's important to recognise that these parasites are host-specific – in the case of the squirrel, it is the chosen host and the fleas need the squirrels in order to survive, and are not therefore about to choose you instead. Parasites, though unpleasant, are also fascinating. Squirrels have a range that often includes ticks, lice and mites, and during the moulting season – twice a year, in spring and autumn – these unwanted passengers cause a considerable amount of itching and discomfort. By frequently changing and moving their dreys, a squirrel helps minimise the problems of parasitic infestation.

During the spring moult, a squirrel can look very shabby, and after the extended winter spell in the drey, parasites have often taken hold. Now there is frequently considerable scratching, exacerbating the hair loss happening already because of the heavy moult. During the spring, I have noted that squirrels appear to lose their coats from the front to the back, leaving the animal looking as if it is only wearing breeches. In the autumn, the reverse happens, so that it seems as if the afflicted only has a waistcoat on. If Beatrix Potter could have seen these individuals, I cannot help thinking she would have created some priceless artwork based on the seasonal moult patterns. In spring, most adults lose their fabulous ear tufts too. This may temporarily alter their appearance, and it is now, perhaps, that their enchanting little faces do become rather ratty.

The moult can be drastic, especially if there is a parasite burden too, leaving the squirrel with large, unsightly bald patches. I receive calls from concerned people who have a particularly balding squirrel in their garden asking what can be done to help. Unfortunately, there is little that can be done to ease the itching and scratching caused by fleas or mites, though we have devised a method of putting a specialist small rodent anti-mite product onto a little piece of sponge attached to the lid of the peanut feeder boxes. When the squirrel goes in, it rubs on its shoulder blades. This has worked well in extreme cases, but it is not widely done, as far as I know, and has to be carried out with great caution, and under the auspices of a qualified vet. Of course, all wild animals and birds go through the stress of the moult. If you think of a cock blackbird in spring when he is in the peak of health, his gleaming coal-black plumage looks as though it has been covered with glossy varnish, there is not a feather out of place and his eye rings shine daffodil yellow. As summer advances he becomes a drab fellow, his finery is temporarily gone for he is exhausted from the demands of rearing several broods of young. He too is balding, and is in heavy moult. Now he keeps a low profile, almost as if he is ashamed. And when deer go through their moult, they look as if they're wearing moth-eaten charity shop fur coats until their lovely new gleaming red coats come in. Moulting is a necessary process, and few creatures look their best at this time, least of all squirrels.

A squirrel becomes sexually mature at about a year old. However, on occasions, particularly when there is an abundance of food, kits born very early in the year may breed in their first year. However, this is more unusual in Scotland. Mating takes place throughout the year, with two peak seasons – one very early in spring, and a second in midsummer. When food is abundant and conditions are right, a mature squirrel may give birth to two litters in a year. The gestation period is approximately thirty-eight days, and most spring litters are born in March and April, with subsequent litters born later in the summer, though there are exceptions. Litters have been recorded in January in parts of England where the weather is milder.

A squirrel's courtship is not dissimilar to its eating habits – noisy. They are promiscuous and there is no loyalty between a pair – unlike some birds, they do not forge a strong pair bond. Their only reason for coming together is to mate. When the time is right, male squirrels seek out a female and start to give chase. To begin with this is usually unappreciated, and the eager male is rebuffed and often scolded in the process. Then a mad chase will begin accompanied by chattering vocalisations, and the sound of claws gripping fast to bark. Often other males appear and join the mêlée as the chase becomes ever more crazed, particularly if foreplay is centred around a large tree trunk. Watching these antics it soon becomes impossible to work out who is who, and the watcher can feel quite giddy due to the speed of the squirrels' descents. I have frequently watched courtships – often from our kitchen windows – amid showers of bark flakes and chitter-chatter, yet I have never seen mating taking place. I find myself glued as another mad ritual begins. Sometimes the games continue for days until one male is successful. After mating, even if she is not pregnant, the female will not usually accept further advances from other males until the next time she is in oestrus.

Once their gene-spreading work is done, male squirrels have nothing further to do with the breeding business. It is the female who will construct herself a substantial nursery, or refurbish one that she has already used, and it is she who will raise the young squirrels entirely alone.

Litters usually consist of an average of three to five kits, but sometimes six or seven have been recorded. Like baby rabbits, the kits are born naked, blind and wholly dependent on their mother. Now she won't venture far as they require her rich milk at regular intervals, and also rely on her warmth to keep them from the danger of chilling. Kits don't open their eyes until they are around three weeks of age, by which time a soft, velvety ginger fuzz spreads over their little bodies like a covering of mould, and their long, tails start to look more like those of a squirrel, rather than those of a rat. It is now that there is the biggest disparity between their huge hind feet and their tiny bodies. Claws are also well developed from an early age to enable the inexperienced youngsters to retain their crampon-like grip once they take their first inquisitive steps out of their dreys at around eight weeks old. Their mother continues to suckle them at intervals for nearly three months.

Watching from our kitchen window, our excitement mounts at this stage, when the mother leaves her hidden litter and comes into the feeders more frequently. We can see that she is lactating due to her swollen mammary glands and more prominent nipples, and hope that soon her young will also put in an appearance in the garden. Once the kits stop suckling she regains her neat, svelte shape almost immediately. If only it were that easy for us sugar-addicted humans!

Young squirrels may be seen playing in close proximity to their drey, though with dense leaf cover a good view is never guaranteed. Although squirrels are not territorial, they do appear to like to protect their patch and at this time the resident adults may chase the youngsters in an endeavour to push them away from the area.

Once weaned and outwith the safety of the natal drey, baby squirrels must quickly become independent. They frequently struggle to find enough food, and it is now that they easily succumb to starvation, and are also at risk of becoming severely chilled in our increasingly heavy summer deluges. Their first attempts at drey-building may be poor, and in some cases probably don't provide proper protection from the elements. However, when the habitat and cover are right, and there is plenty of food, they will of course stand a far better chance. It is in summer that food put out by garden owners can provide a lifeline for these half-grown youngsters. Mortality rates at this time are high, but a regular food supply put out by a kind gardener can make the difference between life and death.

6

A Grey Area

In his book *The Naturalized Animals of Britain and Ireland*, Christopher Lever writes: 'Humans are inveterate and incorrigible meddlers, never content to leave anything as they find it but always seeking to alter and – as they see it – to improve. In no fields is this truer than in those of the animal and plant kingdoms.'

Probably one of the best, or should that be the worst, examples of a misguided introduction to the British Isles is the North American grey squirrel. The saga of the larger grey in our midst is not a happy one. Ever since its arrival, it is fair to say that it has gone forth and literally multiplied. And its pernicious spread seems increasingly likely to be a problem we will probably never resolve.

The first clear evidence of the grey squirrel's arrival comes from Henbury Park in Cheshire where T.V. Brocklehurst released some in 1876. However, they may have been released in Kent even earlier, around 1850. Prolific breeders, the new arrivals swiftly established themselves, aided by numerous landowners who followed the trend by introducing grey squirrels on to their land too. Reflecting on it now, it seems ironic given that in many places around the British Isles red squirrels were being culled: why did the Victorians feel it was a sensible idea to import more squirrels? There was a continual demand for grey introduction, as it became popular and was regarded as the thing to do, something new and 'frightfully fashionable', a quirky animal to adorn the countryside.

In 1889 G.S. Page of New Jersey imported squirrels for release in London's aptly named Bushy Park – however, this was not a success, though they are found all around London's numerous parks and green spaces today. Conversely, those released around the same time at Woburn Abbey in Bedfordshire had no problems proliferating. And thus, like a wildfire wind-driven on a tinder-dry hillside, this large North American rodent began its rampant spread across the countryside at an alarming rate, through no fault of its own.

The first grey squirrels in Scotland were released at Finnart on Loch Long, Dunbartonshire, in 1892 – again by G.S. Page of New Jersey. Their colonisation here was equally impressive, and during the next twenty-five years they were found over a 300-mile radius. There were several other Scottish introductions. Grey squirrels kept at Edinburgh Zoo escaped in 1913, and it is likely that the abundant grey squirrels that continue to dominate the city's parks and gardens stem from those first escapologists.

As the new squirrels began to establish, it quickly became clear that where they shared the same woodlands as the resident red squirrel, the native population quickly dwindled. Rumours spread as people blamed the grey for literally killing the red. By 1938 it was declared illegal to import any more greys, or even to keep them in captivity. However, it was too late. The damage had been done.

At the start of the 1950s, the Ministry of Agriculture, Fisheries and Food offered a bounty of one shilling per grey squirrel, and by 1958 an estimated £100,000 had been paid out, but there still seemed little decline.

Many people remain under the misconception that the grey squirrel kills the red one. It does, but not in the literal sense. It is its mere presence that inadvertently leads to the displacement and subsequent loss of the red squirrel, through lack of food availability. It is fair to summarise by saying that where the grey proliferates, the red retreats. It's a complex issue that is perhaps best understood when we take a closer look at the grey.

Compared to its red relation, it is almost twice the size and weight. Though it may often have ginger colouration to its pelt, almost as if it has henna highlights, it is not easy to confuse the two species. It is also noteworthy that the grey squirrel has no ear tufts and its tail is usually less bushy and luxuriant. It is far more robust and is therefore less susceptible to disease. And, unlike the red squirrel, it is not habitat-specific and can thrive in higher numbers in smaller isolated pockets of deciduous or coniferous woodland. The grey consumes many natural foodstuffs, including seeds and acorns, long before they are ripe enough for the delicate digestive system of the red. In short, when food is at a premium and the two species exist side by side, the grey has the ability to outcompete, and therefore comes out on top. The grey is also less sensitive to disturbance and has adapted particularly well to living in close proximity to man in town parks and public gardens, where it freely takes advantage of an eclectic diet of titbits, and where it spreads its grey genes with impunity. We are lucky that the two species do not interbreed, so unlike the issue surrounding red deer and the non-native Japanese Sika deer – also introduced to this country – the two species do not hybridise; if they did, it could lead to the loss of a particularly unique and special little rodent.

However, there is a far greater horror lurking unseen in the grey squirrel population. An estimated 70 per cent carry the lethal squirrel parapox virus (SQPV). Though the carrier is immune to this, squirrel pox will swiftly wipe out an entire population of reds in little more than a fortnight. At the start of the 1900s an unknown virus began to kill entire populations of red squirrels in various parts of England, but it was not until the early 1960s that this was recognised by vets as a pox virus. By early 2000 it was confirmed that the affliction was fatal to red squirrels but did not affect greys, and a few years later they were found to be the carriers. The first red squirrel with the virus in Scotland was discovered in 2007, since which time there have been sporadic, worrying outbreaks, particularly in southern Scotland. Those that are afflicted are pathetic and look not dissimilar to rabbits with myxomatosis, with swollen eyes, skin lesions and ulcers. There are usually few survivors. Scientists at the Moredun Research Institute in Edinburgh have been working to develop a vaccine since 2009, but administering this to wild squirrel populations will not only prove difficult, but also extremely costly. We have to hope that they are successful and ways can be found to address the problem.

What began as a trend to introduce an exotic species to parks, woodlands and gardens is surely a catastrophe for the red squirrel. Sadly, this is the case with most, if not all, alien additions to our flora and fauna. Few appear to raise the issue of the introduction of some 40 million non-native pheasants and red-legged partridges (the latter with Mediterranean origins) into the wild every year for game shooting purposes. However, there can be little doubt that they too have a serious impact on our ecosystems and consume vast amounts of invertebrates and natural food to the detriment of native birds and mammals, particularly hedgehogs, even though game managers do put out supplementary feeding that doubtless helps some other birds, particularly seed eaters. The crash in numbers of our only poisonous snake, the beautiful adder, has recently been attributed to the high numbers of pheasants, which consume large numbers of reptiles, including young adders. We are currently facing a biodiversity crisis. Food sources are not infinite, though advocates for pheasants could argue that the pheasants themselves provide a food source for predators including foxes, badgers and larger birds of prey. Whatever the argument, it is surely important to recognise that there will always be serious repercussions when a non-native species is set free in the wild.

It seems that wealthy Victorians were passionate about trying out new animals, birds and plants, and they also liked to adorn their big houses with glass cases full of dramatic taxidermy specimens. If they found something unusual or particularly beautiful, even a new songster they had perhaps not seen before, then there was a high possibility it would be shot and stuffed. And there is a certain irony to the fact that when they went back to have another look the next year, or the next, they failed to understand why whatever it was that had taken their fancy the year before had now vanished. Nothing was safe. Feathers and skins provided adornments for elegant ladies viewed as the ultimate fashion icons. The millinery trade, with its outrageous feathered hats, was responsible for pushing numerous bird species to the brink, none more so than beautiful members of the grebe family, as well as some seabirds, including the smallest and prettiest member of the gull family, the kittiwake. In fact, it was in direct response to this grim trade that in 1889 Emily Williamson founded the Society for the Protection of Birds – now the RSPB – because she was so concerned about the destruction of bird populations as a result of the barbarous trade in plumes for the fashion industry.

Taxidermists not only worked on specimens to reproduce them true to form, but there was also a fad for rogue taxidermy. This technique used various body parts from different animals and birds to create macabre conglomerations. Red and grey squirrels featured heavily – some were transformed with birds' wings into 'flying squirrels'. Others might be dressed up in tweeds or elegant clothing, or glorified with feathered headdresses. One horrendous example I had the misfortune to see in an English museum had fangs, birds' feet instead of forepaws, and roe deer antlers. Rogue taxidermy was a repellent idea and ran parallel with the Victorians' misplaced judgement in bringing species to the country that put our native fauna and flora at risk.

Though the red squirrel will strip bark too, the damage it does is not thought to be very serious. Conversely, the grey is seen as a wrecker of young trees. Vigorous ring-barking (when bark is stripped from the entire circumference of a trunk) can lead to the loss of numerous saplings in a given area, and in cases where the bark is partially stripped, loss of the tree's resilience and vigour is another problem. Frequently afflicted trees will die or remain stunted. Therefore the grey squirrel's presence is understandably unappreciated by foresters and woodland managers. Scientists believe that large numbers of greys in woodland can also disrupt important, symbiotic associations between plants, fungi and invertebrates. Many gardeners who put out food for birds complain too about the dominance of the grey squirrel at feeders. They are tenaciously smart mammals, and even the most sophisticated baffles and anti-grey devices and deterrents prove ineffectual. What may start as amusing entertainment as the thwarted grey squirrel deals with slippery poles, tightropes, mesh protection boxes and similarly Heath Robinson contraptions, soon turns into an annoyance that many gardeners won't tolerate. The squirrel usually wins. Working out the best method to reach the food appears to be something a grey squirrel enjoys.

As the dreadful truth of having an alien species in our midst unravelled, a new war began in an endeavour to eradicate the British countryside of another of our foolish follies. Poisoning with warfarin has been used, but it is indiscriminate and accounts also for the death of numerous non-target species; even so, on occasion this method is still employed. Grey squirrels have no legal protection under the Wildlife and Countryside Act of 1981, and it is illegal to release any taken to wildlife hospitals and rehabilitation centres. Unless they are kept in a cage, which is still permissible, then – controversially – the only option is humane despatch. And this is not to everyone's liking.

Greys are live-trapped and humanely killed in many areas, particularly hotspots for red squirrels. Understandably, for some people this is extremely upsetting, and where the grey is the only resident squirrel, there has been uproar. I am fortunate to live in one of the best areas in the entire country for red squirrels, and at the time of writing, the appearance of a grey is an extreme rarity, but they are moving closer. However, I sympathise with those who live in the urban environment and for whom the cheery, opportunist grey squirrel can prove a lifeline – a creature they see every day, perhaps their only connection with wild nature. It is hard to accept that an animal that is here through no fault of its own should now face the death penalty. It reminds me of a story about the hated rat, written by a man who had been imprisoned in squalid conditions in a foreign jail – with a rat his only companion. He wrote movingly about his relationship with a rodent we are conditioned to detest, but which he found to be not only a very meticulous animal, but also surprisingly intelligent. In his case, it proved to be a lifesaver and its visits to his cell helped to keep his spirits up. Nature is vital to our wellbeing, and watching and feeding grey squirrels may provide a lonely or anxious person with solace and that all-important sense of connection.

When in my late teens I briefly lived in London, and I vividly remember the antics of the parks' numerous grey squirrels, the pleasure they gave me, and the importance of those moments spent in their company. It is proven that time spent with nature is indeed beneficial to us all and can help with mental health disorders, anxiety, depression and stress, something that really came to the fore in 2020 when the entire country was under lockdown. For some, a grey squirrel may be the bridge between stability and breakdown.

People frequently ask me, 'When you love animals, how can you be so heartless and advocate the removal of the grey squirrel?' And then another debate will commence. I accept that the idea of mass culls is abhorrent. After all, the grey squirrel is blameless and did not choose to be in our midst. So how do I reply? I have always believed in being honest about wildlife conflicts. The hard fact is that if we want to protect our native red squirrel, then we will have to continue trapping and removing the grey, especially in areas where they infringe on prime red squirrel habitat.

It is likely that we will never now be free of the grey in this country. Some say that we should instead accept it as part of our ecosystem, in a similar way to the current situation with introduced ring-necked parakeets, which have become a dominant species in most London parks, and the small muntjac deer too. There are also schools of thought that say removal of greys is a waste of time and does not help the red. I am not sure that I agree with this. For the red squirrels living in Caledonian pine forests and healthy mixed woodlands in strongholds such as Speyside, Perthshire, Angus, Aberdeenshire, Argyllshire, and Dumfries and Galloway, and on the islands of Arran and Anglesey, where there are also thriving populations, keeping the grey out is of the utmost importance.

7

The Compost Café

My desire to work closely with animals began early, and since that time I have looked after a wide and varied number of species, including quite a few exotics. What started as a sideline, or perhaps I could call it a hobby, has become more of a way of life coupled with my work as a writer and photographer. Wildlife rehabilitation provides me with a constant source of fascination (as well as worry) and a never-ending learning journey in which I realise that the more I learn, the more I realise how little I actually know.

Animals have always surrounded me. From as far back as I can remember I have always had a desire to become a vet. Not only was I an animal-mad tomboy, I was also a bookworm. Books can have a major influence on the direction of our lives, and three authors in particular left their indelible mark. While Beatrix Potter will always remain my favourite author for early childhood, hotly followed by Alison Uttley's *Little Grey Rabbit* books, the next writer who captivated my imagination when I was a little older was Hugh Lofting, with his Dr Dolittle series. Dr Dolittle was a great character, but not only could he cure animals, he could also communicate with them. When I was about ten years old, I began to try to model myself on him. I wanted to talk to animals too. I should add a note here: while the books were sublime, the same cannot be said for the latest film versions of these brilliant tales. They are not only inaccurate, but the Americanisation of such a typically English gentleman as Dr Dolittle seems to me to be an insult. Rex Harrison played the part perfectly in the first black and white film version in 1967. While his relationship with the animals in the film besotted me, handsome Rex Harrison besotted my grandmother. Thereafter it was downhill all the way. Hugh Lofting would definitely have been unhappy to see how the new films veered totally away from his books.

Next there was James Herriot and I consumed his books too, and when we were farming full-time I found myself constantly able to recognise various ailments and afflictions simply because his books were so informative, as well as captivatingly entertaining and humorous. Though it soon became blindingly obvious that my brain was sadly lacking in the scientific department, and training to be a vet was simply not going to happen, these wonderful stories made me more and more determined to make a life working with animals. I have always loved hand-rearing orphans, treating sick animals and birds, and find all things relating to veterinary matters fascinating. And that includes parasites!

When I was in my early teens my mother and stepfather, Mike, moved from Ardnamurchan to a farm in Angus, and then eventually to a beautiful property in Kincardineshire. Mike was to have a major influence on my adult life, and was also responsible for providing me with some incredible opportunities, all of which have helped to lead me on my current path. Mum had visited a friend near Perth who had dozens of red squirrels coming into her vast, sprawling woodland garden. Mum never stopped talking about it. Their new home in Kincardineshire was surrounded by mature beech woodland and close to small pockets of native pine forest in the glen foothills that helped to provide ideal, if somewhat fragmented, red squirrel habitat. Occasionally a squirrel or two put in a brief appearance in the garden. Mum set about attracting more.

My mother loved animals too, and for many years she bred exotic, rare breed macaws and cockatoos, including critically endangered scarlet and hyacinth macaws. Her lifelong passion for parrots probably has a bearing on why I am called Polly, but then again she was also pretty partial to putting kettles on and making tea. She taught me much about the importance of good animal husbandry. She never did things by halves: if she was going to do something, she did it exceedingly well. Where the parrots were concerned, if they could not be free in the wild, then they must have the best. She had state-of-the-art aviaries made, complete with a feed room, a small kitchen area with sink and microwave for thawing out frozen food for the birds, particularly corn on the cob. This she bought in bulk because the parrots loved it. Those birds had as fine a diet as any parrot could wish for. Birds and animals kept in captivity tend to be very wasteful, and all the half-eaten nuts, seeds and fruit (always still fresh) were added to a large compost heap on the edge of the wood. These lush delicacies enticed squirrels from afar, and it seemed that they loved corn on the cob too. Mum welcomed them ecstatically. On the other hand, her welcome for the hordes of rats that also began to appear was less effusive. But when it comes to wildlife you really cannot be choosy. There was a pig farm close to my parents' home, and this was partially blamed for the explosion in rat numbers. However, it didn't take much to work out that it was far more likely to be the fabulous compost heap café that attracted the Kincardineshire gourmet rats, as well as a growing population of jackdaws, and indeed the delighted squirrels. Well, you really couldn't have your cake and eat it, one of her grandiose, patronising friends told her.

Far worse was the fact that the farm also had an ever-growing population of feral cats. Unfortunately they were badly infected with a form of cat flu and were too thin, mangy and lacklustre to do much work in rat removal, but for some reason they were exceedingly adept at catching squirrels. My mother couldn't bear it and, with the farmer's permission, I am afraid to say she successfully removed a large number of them. Sometimes they were trapped and taken to the local vet, never to return, and sometimes she shot them. Anything that put the squirrels at risk was unacceptable.

There was an incident when I was staying with my parents when a cat caught a squirrel as it was eating discarded hazelnuts on the gourmet tip. Mum roared, took off her welly boot and hurled it at the cat. She missed but it took fright and instantly dropped its prey. The victim lay on the leaf mould, rigid and trembling. This was my first experience of handling a red squirrel, but this inauguration into red squirrel care was not a happy one. The casualty was in a severe state of shock, and it was foaming at the mouth – we did not hold out much hope, but we took it into the house and put it safely in a warm, dark place anyway. Its breathing was laboured. It lay motionless for a couple of hours before it died. We were both very upset. I wanted to learn more about what could be done in such circumstances. If we had known more, could we have perhaps saved it?

In my final years at Gordonstoun – the only school I ever enjoyed – for community service I worked with the local vet. It was a wonderful opportunity to learn more about animal medicine and I loved being allowed to assist with operations – usually this involved shaving a lot of cats' tummies ready for them to be sterilised. I would have long chats with the vet, who generously filled me with information, while carefully explaining the minutiae

of what was going on and discussing various infections and afflictions. Occasionally, a few wild casualties were brought in. He was
always adamant about stress – 'it's really the biggest killer of all for
wild things and you must always take this into account when you
are treating a wild casualty'. It was very sound advice. As I was to
find out increasingly often, stress in highly strung animals with a
fast metabolism is very hard to beat indeed. I have since lost many
squirrels, and though the injuries they have sustained might also
be serious, stress will all too frequently be the actual cause of death.

When I was working in a privately owned zoo in Northampton-
shire, caring members of the public delivered a constant stream
of injured and orphaned wildlife; some of the inmates too might
need attention. I was greatly drawn to this first aid and medical side
of the work at the zoo and saw some amazing things while there.
Once, a giant hornbill sustained a serious crack in his massive bill,
and the specialist vet who came to treat him used superglue to
mend the damage. It worked astonishingly well. I was fortunate
to hand-rear two puma cubs whose mother had died, and to assist
with an operation carried out with another ailing puma laid on a
large board positioned over a bath. He had swallowed a soft toy
fed to him by a visiting child. Eventually, after a nerve-racking
hour, the toy was retrieved and the puma stitched back up – I was
intrigued as I peered into the cavernous area where the toy was
lodged.

I also loved to work with the tawny owls, little owls and barn
owls, and the kestrels and buzzards as well as numerous hedgehogs.
Grey squirrels were frequently brought in, but there were no reds
anywhere in that part of the country. The wildlife rehabilitation
and veterinary work with the zoo's other inmates was something I
relished. At dawn, the surrounding broadleaved woodland echoed
to the orchestrations of not only British songbirds, but also an
eclectic percussion of sounds from the rainforests as macaws,

cockatoos, toucans, mynah birds and exotic owls, shrieking monkeys and numerous other foreigners filled the sun-dappled daybreak. I have always been an early riser, but here it was hard to stay in bed much later than 5 a.m. I always volunteered to do the early shift, much to the relief of the other staff, who preferred to stay in bed.

While red squirrels were sadly absent, all the time I was learning, gleaning and collating, filling my head with information on animal husbandry and witnessing first hand what worked and what didn't. I was also learning that every animal and bird needs different handling – some are even more prone to stress than others. While tawny owls are usually fairly easy to deal with when sick or injured, the same cannot be said for sparrowhawks – these raptors are the stress mechanics of the avian kingdom. You only have to look at the nervous, glaring yellow-eyed stare of a sparrowhawk to see it's not a bird that is laid-back. Much later on, when I would begin to work closely with injured and orphaned red squirrels, this was all useful background. One thing was blatantly obvious: the more I worked closely with wildlife, the more I realised that this was going to be my life's vocation. I also realised, importantly, that swapping and sharing information with other rehabilitators was a vital aspect to progression.

In nearly all cases, the casualties that have come to me since have, with very few exceptions, all been caused by human actions. And I have always firmly believed that though so often it is not possible to save or return badly injured animals to the wild, it is important to endeavour to rectify the problems that we have caused.

8

Care of the Wild

Being a substitute for an animal's mother is fraught with problems. While it may seem like an attractive prospect, it demands extreme dedication. The younger the animal or bird, the harder it can be, though that being said, if an orphan is too old, it can take far longer for it to accept and adapt to human intervention, and that brings another batch of problems.

When mammals are born, their mother's first milk – colostrum – is vital to their future health. Packed full of antibodies, it is far richer and thicker than the subsequent milk flow, and helps to protect their immune systems and give them the best chance in life. For a rehabilitator, receiving very young newborn mammals that have probably not had this immediately means the challenges will be far greater. If the animal is more than three days old, you are probably in with a chance of success. If it is younger, then it will be susceptible to all manner of ills. Though it is possible to use colostrum substitute, it will never be as beneficial as the mother's own.

When it comes to wild mammals, it is vital to ensure that the milk substitute being used is correct from the outset, and to try to ensure that it matches the fat content of the natural mother's milk where possible. Milk substitutes have become better and better, though there are some on the market that don't work well and should be avoided because they do not provide necessary vitamins and minerals. Cow's milk is, in nearly all cases, unsatisfactory and will usually lead to bad bouts of diarrhoea. One of the most important points to understand about milk substitutes is that once an orphan is on one variety and is thriving, it's safer to stick to that and not make any changes – changes, and particularly sudden changes, are almost certain to lead to digestive disruption. If it's the only way forward, then introducing the new milk gradually helps to avoid trouble. Diarrhoea, often referred to as 'scour', is debilitating. In order to stop it, the afflicted has to be taken off the milk and instead fed water and electrolytes. Of course, being starving hungry and desperate for more nourishment but still unable to eat solids makes it a very difficult business, as milk can only be gradually reintroduced, and only a weaker solution at that. Baby animals usually cannot tolerate periods of fasting, so the key is to try to avoid the problem at all costs. Bottles and utensils must be kept scrupulously clean, and the animal made as warm and comfortable as possible so that it is not stressed in any way. My heart sinks when any of the animals I am hand-rearing starts to scour, because it usually means if you cannot stop the problem, death won't be far behind. This may sound dramatic but it is a hard fact.

People often ring to say they have found, for example, baby hedgehogs or leverets. They first ask for some advice on how to rear them and tell me that this is something they are very keen to do themselves. Often in less than twenty-four hours the telephone will ring again. 'Can you possibly take them now as we have found that really we are just far too busy and cannot fit in so many feeds, with our work schedule and everything else.' I suspect that what has actually happened is that they have found it too tying, and also the associated mess is too much to contend with. Perhaps I am being cynical, but while the idea is attractive, the reality is extremely labour-intensive. Hand-rearing is also fraught with heartbreak. Just when you think that things are progressing well, you may rise at dawn and find that your baby has succumbed in the night. As I learned with the red squirrel that was attacked by a cat at my parents' Kincardineshire home, stress is the silent killer – it's not always easy to see, and the unnatural situation and loss of the animal's mother compromises the baby to the extent that it can all too readily pick up infections, stress-induced pneumonia being a common culprit. Even when quickly diagnosed, there are no guarantees that it can be overcome. It may seem as if some animals simply have a death wish. There is only one consolation in these circumstances, but it's not much help – the animal was probably already nearly a statistic when it was brought in, and therefore you have at least given it a chance, done your best and could probably not have done more. If patients die in the first few days, it's upsetting and disappointing. What I find far, far harder to accept is the occasions when I have had an animal or bird for a couple of weeks or longer and invested a great deal of time and energy into it, and then it dies unexpectedly quite far down the line, just when I think it is thriving. It's not unusual.

*

While I had long hand-reared numerous wild animals and birds, including a few exotic species while working in the zoo – Arctic foxes, puma cubs, Vietnamese pot-bellied pigs, eagle owlets and a macaque monkey – as well as macaws and cockatoos with my mother, it was quite a long time before the first red squirrel kits were brought to me.

I have always kept wildlife diaries. Some entries are sparse and matter-of-fact, others involve emotion and detail, but all have proved to be a useful record to see not only what I might be able to do better, but also what did actually prove successful. These diaries also reveal another story. It is the story of the decline of our wildlife. There is no getting away from the fact that we are losing species at a terrifying rate. As I get older, I find myself lying awake in the middle of the night worrying about it, and it has continued to fuel my mission to connect more people to the joys of the natural world in the hope that this will make them see how vital it is to protect it.

As a child with a passionate ambition to become a vet, under the guidance of my mother, I had often tried to help needy fledglings, road casualty hedgehogs, indeed any sick and ailing wildlife that we found, but it was not until I was settled on a small farm in Kincardineshire in the mid-1980s that I began to work closely with the Scottish SPCA, taking in a range of casualties. Looking back through my diaries of that era, kestrels, tawny owls and buzzards were my most frequent patients. Compared to many other birds, they usually did well while in captivity and did not tend to suffer from stress to the degree of birds such as sparrowhawks, goshawks or long- or short-eared owls.

Noteworthy is the fact that in the past ten years I have only received one kestrel – they are now an extreme rarity in the area of Highland Perthshire where I live, and it's many years since I last had a short-eared or long-eared owl. I meet people who tell me that this is not the case, that indeed they have just seen, say for example, a kestrel, and that there are plenty of them still around. It's about shifting baseline syndrome and it's hard for someone younger to believe that what they see as plenty of a certain species now is in fact so much less than what was there previously. Unfortunately, it's a dangerous misperception.

When it comes to kestrels, a little raptor I have always adored, I could optimistically hope that the reason I don't receive many any more is because they have not needed help. However, I know this is not the truth. Numbers of kestrels continue to fall and though there are still hotspots, in most areas of the British Isles there is decline among a whole host of species. Reasons for this include a dramatic change in farming policy, with a swing away from 'extensive' agriculture on smaller farms, with a rotational system that left plenty for wildlife, towards intensive agriculture, and the subsequent loss of hedgerows and field margins rich with wildflowers, invertebrates and small mammals. Now more silage is made than hay, and fast-growing heavy cropping grasses are grown instead of old-fashioned meadow mixes, so that several crops may be cut throughout the summer. And of course, everything is smothered in pesticides and fertilisers before gargantuan machines sweep into massive fields to chop all in their wake – and that includes mammals and birds, their nests, eggs and chicks. Kestrels are also victims of our increasingly busy road network as they hunt for small mammals along motorway verges. They have suffered drastically as a result, and while other species, such as the adaptable buzzard, seem to have fared well, there is no doubt that the loss of the 'windhover' from our countryside is a worrying trend and reveals a great deal about the general state of the environment. Like the red squirrel, this beautiful little falcon is less able to adapt if the habitat is not right.

Hedgehogs too have suffered badly for the same reasons, and my wildlife diaries are witness to the drop in numbers that I now receive compared to what I would have had merely three decades ago. Many will blame the explosion in the badger population for the demise of our much-loved prickly-backed urchin and though there can be no denying that the badger is one of few animals that enjoys a hedgehog snack and can extricate the soft body parts from the armoury of spines, these two ancient species have coexisted since the dawn of time. It's a far more complex situation that we will look at later.

Hedgehogs are delightful. Ancient creatures that have had little need to evolve since they first appeared some 1.5 million years ago, they are curiously enchanting. Their hibernation is fascinating, and their ability to roll up into a tight ball is another mysterious facet of their eccentric nature. They hide surprisingly long legs under their pelmet of prickles and are agile climbers, running fast when the need arises. Unusually, this wild mammal lets us get close and doesn't adopt a fight-or-flight reaction like most other wild things. Instead, when frightened it usually curls up until the threat has passed. This reaction is disastrous with traffic.

Hedgehogs regularly visit urban gardens, snuffling around at dusk searching for invertebrates, or appearing at the doorstep to consume food left out for them. And this makes them feel approachable, a part of our lives. Hedgehogs have an omnivorous diet largely based on slugs, snails, worms and other invertebrates, as well as fallen fruit, eggs and carrion, and they relish dog or cat food.

I have been taking in injured and orphaned hedgehogs since I was a child. Litters born late in the season are often too low in weight by the onset of winter and have not stored enough brown fat to carry them through hibernation in the wild. In the autumn, wildlife sanctuaries receive numerous such animals to overwinter. We have made a special hoggery where we house these youngsters in a large safe space in the building on our small farm that was built originally as a piggery. Thickly bedded with dry leaves and leaf litter, hay and moss, once installed they instantly start to make their own nests. Ferrying their chosen bedding in their mouths, they tightly pack it into hedgehog houses. Sometimes half a dozen snooze together, while others prefer to sleep alone. A surprising amount of heat is generated in these safe hibernacula. As night falls they re-emerge huffing and puffing while noisily eating their food. Some are pugilists and barge one another out of the way, spines erect – it almost seems possible to see them frowning crossly. They are enthralling.

Depending on the weather, our winter visitors are released early in spring. I liaise with the Scottish SPCA taking some of their overwintered hedgehogs too. Finding suitable habitat is an increasing problem. It's not so much the worry about badgers, but the need for them to be far away from roads and in the richest areas of native woodland that are not marooned by monocultures and overgrazed deserts.

While release is a high point, it also fills me with concern – every single hedgehog matters more than ever. Many of my overwintered hedgehogs have recently gone to Aigas Field Centre at Beauly, where extensive habitat restoration carried out by the Lister-Kaye family and their excellent team of rangers has brought exemplary results. As I watch their wonderful forms trundling off into the undergrowth, I ask myself – will they make it? Restoration ecology is the solution to the dearth of the hedgehog, as it is for every other living thing, including us. It starts at the bottom, with microscopic soil organisms. If the earth is unhealthy, overworked and filled with chemicals, then little can survive. We need to study the bigger picture. On a positive note, there are many excellent farmers who are working with groups such as the Nature Friendly Farming Network, an organisation that unites farmers who are passionate about wildlife, and helps them to find ways to farm sustainably. The Soil Association is a charity whose activities include campaign work on issues including opposition to intensive farming, support for local purchasing and public edu-cation on nutrition, as well as the certification of organic food. Soil health is key to the success of all life. Sadly, farmers are under enormous pressure to produce more and more; they have been

squeezed into a situation that has greatly contributed to the subsequent catastrophic loss of biodiversity. Leaving areas wild and
stopping the use of chemicals and artificial fertilisers can mean far
lower yields and problems with disease issues that may push them
to the brink of financial collapse. It's a vicious circle governed by
political mayhem and an unsustainable explosion in the human
population. In many areas, we have no space left for wildlife,
and those farmers who work tirelessly to help increasingly need
financial incentives to do so. At times it appears as if our native
wildlife has to pay rent to exist on land that previously provided
food, shelter and a mate.

Like the red squirrel, we didn't always love the hedgehog. On
game-rearing estates hedgehogs were heavily persecuted due to
their habit of stealing the eggs of ground-nesting birds, and frequently ended up on a gamekeeper's gibbet alongside a host of
other unfortunates already mentioned and classified as 'vermin'.
Vermin is a word I hate. Its definition in Chambers English dic-
tionary is: 'a collective name for obnoxious insects such as bugs,
fleas and lice, animals such as mice, rats, animals destructive to
game such as weasels, polecats, also hawks and owls; odious, despicable people: any one species or individual of these'.

Vermin used to describe odious, despicable people seems highly
appropriate, but as for the rest . . . I leave you to decide.

9

Drey in a Lunch Bag

The first time I received orphaned red squirrels to hand-rear is scored too vividly on my memory: 13 March 1996 is a day I will never, ever forget. So traumatic and tragic, so utterly shocking and affecting, that even now, writing about it twenty-six years later, sends a frisson of horror and sadness through me.

I am someone who quite enjoys the winter, but March is a month when most wildlife is at a low ebb, and so are we. The weather can be cruel and relentless. Having survived through the winter, it is this month, with its fickle nature and interminable cold and wet that can push animals and birds to the edge. It's also the time of year that many species are already breeding, only to be set back by late heavy snowfalls and a lack of food.

I was living near Crieff in Perthshire and regularly receiving wildlife from our local Scottish SPCA inspector, Don Wilson. My diaries at this time reveal a wide range of casualties. The entry for that life-changing day reads:

> Don Wilson brought me two orphan red squirrels felled
> in their drey by a woodcutter near Almondbank, Perth.

The woodcutter had sensibly put the entire drey into his lunch bag to keep it intact, and the contents safe.

Then below, there is a paragraph as follows:

Today the most appalling nightmare that could ever happen took place when a crazed gunman burst into Dunblane Primary School, shooting sixteen children dead and critically wounding fifteen others. Their teacher was also killed. The world is left totally numb by this evil, wicked atrocity – nothing else seems to matter any more. Tragic, tragic parents waited outside the school for hours not knowing whether their children were dead or alive. The innocent little ones were all five-year-olds. The gunman then shot himself. As the day wore on more and more horror came to light about this evil, incomprehensible act. God help the world, such inexplicable wickedness is terrifying. And here right on the doorstep in rural Perthshire – a scene of utter devastation, a mass slaughter akin to petrifying gun scenes witnessed in America. How could this happen here, here in Perthshire? I have never felt so frightened and shocked. My heart bleeds for Dunblane.

Everyone's heart across the world bled for Dunblane.

It seems mundane now to write about those first squirrels, and indeed, my diary entries for the time that the squirrels were with me are dominated by the unfolding events of Dunblane. It was impossible to accept and comprehend that this mass shooting – the worst in the history of the British Isles – had taken place only miles from my home, in a quiet Scottish town with a beautiful cathedral where I had numerous friends, many who had children in that very school.

I vividly remember many things about that day – firstly, that I was in Crieff at the jeweller having a new watch battery fitted. The weather was typical of March, and we had had a heavy, wet fall of snow a few days previously. The associated grey sludge dripped and clicked off gutters and seeped down road gulleys and lay on road and pavement like a dead grey beast. It rained heavily, making it treacherous underfoot. Shoppers were hunched, dank and miserable hurrying down the street, bent double against the gloom. I was standing at the counter chatting about nothing in particular when the door of the jeweller's little shop on the town's High Street blew open and a man fell in. He was obviously a close friend of the shop owner. He was ashen-faced and told us that he had just driven up past Dunblane. Already news was breaking regarding the tragedy – he had heard it on his lorry radio. Then all traffic was stopped and he had seen and heard the frightening squeals of sirens as ambulances, police and emergency services roared into the town. He stood there shaking while he regaled us with the unfolding gist of what he understood had taken place. Until that moment we had been oblivious. The three of us stood quaking in utter disbelief, tears pouring down our faces. An engulfing bitter terror threatened to strangle me. I rushed out of the shop and went straight to my son Freddy's primary school where other parents had also congregated, all of us terrified and desperate to take our children safely home after hearing the breaking news.

Once in the embracing warmth of our kitchen, we took comfort by focusing on the squirrel kits. I carefully opened the tightly wedged drey and found inside two squirming ginger babies entwined around one another. They seemed undamaged by the fall. The interior of the drey was densely lined with soft sheep's wool, matted lichens and feathers. It was also full of fleas.

Knitted woollen or fleece beanie hats make ideal substitute squirrel dreys. These can then be placed on a hot water bottle.

Squirrels love comfort and babies settle well inside the womb-like confines of these substitute nests – they appear to feel safe. I estimated the little squirrels to be nearly five weeks old. Their eyes were fully open and they were wriggly, mobile and fully furred. There was one of each sex. To begin with I fed them Lectade (an electrolyte mix that combats the effects of dehydration) from a syringe. They were very thirsty and surprised me by taking it readily, their little front paws with their noticeably elongated digits pummelling the air as if they were stimulating their mother's mammary glands to aid milk flow. I then gradually introduced a weak mixture of kitten milk substitute and fed them every four hours. They took the milk from the syringe but did not take it so readily when I attached a tiny teat. They were almost ready for weaning.

My wildlife diary continues with pages of outpourings and details from the Dunblane massacre and in among it is the unfolding story of the two little red squirrels that also consumed me, that offered a salve for the misery that swamped us all. Red squirrels are joyous little mammals and that is one of many reasons why they are so popular.

After a couple of days I had weaned the babies on to a sloppy milky mixture of liquidised apple, digestive biscuit, nuts and seeds. They greedily took this from a small silver teaspoon with a pointed end. You cannot bend stainless steel but silver can be easily shaped. The spoon had been made into a neat point that was far easier for them to feed from. We joked that these really were babies reared with a silver spoon.

Within a couple of days they were both feeding well from a dish, though the female was always ahead and seemed brighter in every way. During the weaning process squirrel babies get themselves plastered with the sticky, sloppy mixture as they paddle around in their food and at the end of each feed they need to be cleaned to avoid the mixture hardening on their beautiful coats. They don't seem to mind this indignity. The precocious female loved it and emitted soft, contented squeaks.

It does not take many days on the soft milky mixture before kits start to test out solid food. In the wild they would not have this interim period on a soft mixture but would be out nibbling various new things, having an occasional feed from their mother for a short time before they completely leave the drey. They snatched the slices of apple and carrot that I gave them and rushed off to a corner of the box where they sat up to nibble them, not wanting the other squirrel to pinch it. Squirrels are always protective of their food and jealously guard it if there are other squirrels around. Pieces of sugar-free Farley's Rusk are also ideal for them at the start and, as with human babies, are always popular for testing out new teeth.

18.3.96

Weight 4 oz

Both doing well and now eating small amounts of chopped sunflower, apple, carrot, chopped nuts and digestive biscuit mixed. Eating milky mixture every few hours and running up the weldmesh front of their box. They look quite content but still spend a lot of time asleep in their woolly hat. They are enchanting. They had so many fleas on arrival but there is now no sign of these. Probably on us!

With regard to fleas, as I mentioned earlier, these are host-specific. Though baby animals often arrive with a heavy infestation, I don't use flea powder on them. The reason for this is that they are already compromised and stressed having lost their mother. To use a toxic product on them at this stage could be enough to kill them, and so I always try to avoid it where possible. Usually, as my diary suggests here, they seem to disappear elsewhere. But where? None of us have ever been infested.

20.03.96

The babies ate less today when hand-fed and have become more independent. They are fantastic to watch as they play together. They went mad with visible excitement when I gave them each a whole hazelnut, grabbing it from me gleefully and racing off to the corner of the box to sit up holding it in their front paws while chomping loudly. When we watched them doing this, Freddy, who like me was raised on Beatrix Potter, exclaimed, 'Mummy aren't they exactly like Squirrel Nutkin? Look at the way they are holding the nuts!' Small pieces of carrot are also relished in the same way. They now need a bigger box with more space in which to play.

At this stage squirrel development seems even swifter. In the wild, once out of their natal drey they have to be independent almost immediately. This is a very tough time for a young squirrel and is the point where many fall by the wayside and die of starvation. It's also why on rare occasions young squirrels out in a wood on footpaths and in public places sometimes seem very tame and will approach walkers. They are literally so frantic with hunger that they will risk getting too close to people.

A larger box lined with Scots pine branches gave the squirrels far more space and scope to hone their skills, and we spent hours and hours watching their mad antics. March weather continued typically with a mixture of bright sunshine and dour, cold wet. The Scottish SPCA inspector continued to bring me wildlife – a buzzard with a damaged wing from Crianlarich, and another from

Crieff that had flown into a wire. The local vets asked if I would take a mallard that had been hit on the road. They had checked her over, diagnosed severe bruising, but thought she needed some time to recuperate prior to release. To thank them for their concerned ministrations she laid an egg in the surgery. And then it seemed to be a year of tawny owlets and by the end of March I already had seven.

Freddy had a couple of friends to play one day and it led to near-disaster with the squirrels. One of the children was a rather independent, and strange little boy, and he took himself off to look at a book inside while we were all outside digging the vegetable patch. It wasn't until nightfall that I made the awful discovery that one of the squirrels had vanished from the box. I hadn't been in since morning when they were both there; it was obvious that the visitor was probably the culprit. My heart sank. It seemed as if that was the end of it as the back door had been open at the time. We were sure the frightened squirrel must have rushed outside and fled. Of course, the wretched child denied it. The remaining squirrel – the female – was visibly agitated and didn't want her food. My heart sank again. They were both so vulnerable, and I cursed myself for letting the little boy go into the house by himself.

Next morning, a very small, famished squirrel arrived at the back door and narrowly avoided further disaster as it managed to evade the jaws of our surprised collies. The squirrel raced off up the nearest ash tree, where it watched me from a safe distance. But he wouldn't come down when I tried to lure him with a handful of favourite food. I then set a live trap and bated it with hazelnuts. Miraculously, within an hour he was safely caught and duly re-united with his sibling. Together they raced around the box in an explosive game before crashing out fast asleep in their hat.

Over the next couple of months I learned the hard way that squirrels, like badgers (and I had learned this when working in the zoo), are incredible escapologists. While badgers can extricate themselves from most compounds, squirrels seem brilliant at finding minuscule holes that you feel not even an anorexic mouse could wheedle its way through.

Then it was time to move the squirrels out to the big aviary. This was lined throughout with pine branches, the floor surface covered with leaves and pieces of moss and lichen where I hid hazelnuts and cones. With a range of structures to climb on, and suitable places to make nests, it was a squirrels' adventure playground. To begin with their explorations were tentative and then, given more space, they erupted into a fast game of chase until I was giddy watching. Only three days had passed when I noticed that one of them had vanished. It was the male again!

It took a long time to find a tiny gap in a join in the aviary's weldmesh. It really did not look big enough for a breakout, but squirrels are tenacious. Once more I set the live trap and two days later caught the ravenous Houdini, and once again reunited him with his sister. Prior to release both the squirrels needed to learn how to use nut box feeders so when eventually set free they would have a back-up in the big arboretum where I had made arrangements to take them.

Though there were a few red squirrels in the area where I lived near Crieff, there were also a great many grey squirrels around. It was not a good place to release the duo. This was why I had chosen another site where there was a healthy population of red squirrels and no grey squirrels close by. It was also a place where food was always put out for squirrels. Considered release is an integral part of successful rehabilitation, and it is vital that youngsters have access to a ready food supply at this point, and indeed long into the future, to give them the best chance of survival. I put some blue sheep-marker spray dye on the squirrels' tail tips prior to release and for the next three months before it moulted out there were positive reports. The success of their return to the wild was a boost, and luckily the mistakes I made had also taught me that when it comes to squirrel housing, next time I would need to go round all the joins in the weldmesh with a magnifying glass.

10

For the Love of Trees

In 1999 I found myself looking for a place to move to that I could enhance for wildlife. It needed to have some land where I could plant trees and hedgerows and thereafter largely let nature manage nature. Having farmed for much of my life, and having a great interest in native Scottish farm livestock, I had continued to keep a few sheep and for a few years bred Shetlands – not the Flock Book variety, but some whose relatives came off Shetland's most remote and westerly island of Foula. Diminutive, hardy and as swift as the savage gales that sweep across this vertiginous cliff-girt landscape, Foula Shetlands are indeed different, and they are also a variety of extraordinary colours. And all are unique in character. Contrary to the much-stated misconception, these primitive northern short-tailed sheep are far from stupid. So the property I was searching for needed areas of grass too for the sheep, and a place to build aviaries for my wildlife casualties.

I have now had our south-facing smallholding in Highland Perthshire for twenty-one years, and during that time it has changed out of all recognition. I met Iomhair (pronounced Eva, a Gaelic translation of Edward) in Islay during a working holiday there and he moved in with me twelve years ago and since then has done an enormous amount of work too to encourage wildlife. He is as passionate about it as I am.

An old Greek proverb says, 'A society grows great when old men plant trees whose shade they know they will never sit in.' Well, in the two decades since I began a rigorous planting regime, our trees are already reaching for the skies, so much so in some areas that the views of distant Loch Tay and the valley floor have been obliterated. You can't have everything, can you? However, the return of the wildlife has been better than we could ever have imagined.

For as long as I can remember, I have wanted to create a wildlife haven. When I first came here to view the place with my late stepfather, Mike, we knew it had enormous potential, but I was also considering two other properties – one of them high above Loch Ness. It had far more land, and the cottage had been newly renovated. I was particularly attracted to it because it also had a Woodland Grant Scheme up and running, many native trees had already been planted and there was potential for planting more and expanding this habitat-enhancing project. There were also red squirrels in the area. The original croft-sized house had a small extension and there would have been little or no work required before Freddy and I could move in. It lay up a long, bumpy track, where it nestled just beneath the hilltop. There is no doubt it would have been cut off and isolated during bad winters. To me that was part of its appeal. I have never minded long periods of weather-related stranding.

The second property was in the Loch Lomond National Park, at Balquhidder. A tiny cottage in the heart of a suppurating bog was going to prove limiting, but I loved it. Mike had been with me to see it and claimed that the amount of frogs we had seen, and the red squirrels in the pinewoods nearby, largely swayed my interest in this particular place. 'You cannot be influenced by a

puckle frogs,' he said. Once I let my head stop ruling my heart, I
was forced to agree with him. Amphibians are another of my many
wildlife passions. Wherever my son Freddy and I went, it had to
be to a place that I could turn over to nature while also hopefully
restoring valuable habitat. 'Your pet sheep would get bad foot rot
living on that bog at Balquhidder,' Mike laughed later as he pen-
sively puffed on another cigarette. So we ruled that one out. But I
was still drawn to the property near Loch Ness. 'It's far too cut off
and remote,' he said, 'and too far away from your mother and me.'
But we needed good imaginations for the property in Highland
Perthshire. The steep ground had been severely overgrazed and
was as threadbare as a worn-out jumper. The few remaining trees
had been ring-barked by sheep and deer, and had been left like
collapsing skeletons. The fences were worse than a row of rotten
teeth, and were propped up with bedsteads knitted together with
vicious rusting barbed wire and baler twine, buckets and corru-
gated tin. It was the sort of place where a cut from old wire might
easily lead to tetanus. There were more thistles and docks than
grass, and though it was in tidy heaps, there was also a great deal of
agricultural rubbish kicking about, to say nothing about the fact
that the house needed gutting. 'This is the place for you,' Mike
said as he continued to puff. 'It's really got potential. You could
transform it.' But I wasn't so sure. 'You just need vision,' he added.
But vision also costs money, and that worried me. I really didn't
think I could face it, or afford it.

I miss Mike terribly for he was always positive and helpful, and
knew a thing or two about potential. After several more visits and
a great deal of deliberation on my part, I leapt off the precipice
and bought the Perthshire smallholding – grandly called a farm.
On arrival, there was a brutal phase of clearing up decades of other
people's rubbish. It is something I have had to do everywhere I
go. And though it is extremely rewarding, it's also relentlessly

exhausting. There were skiploads of it, and all the fences had to be pulled out and replaced before we could do anything.

I have always loved woods and trees. I have also always been aware that they are the most significant plants on Earth. Their potential to store and absorb the carbon emissions that are driving global heating is mind-blowing. Yet more and more forest destruction will have taken place across the world even by the time you read this. And, devastatingly, the British Isles have a long history of woodland loss. As I mentioned earlier, less than 13 per cent of this country is wooded.

Our land here on the farm is steep; the fields form a sunny punchbowl, and given Perthshire's high rainfall, I knew trees would thrive. There were a few red squirrels in the surrounding woodland, but none ever came to the farm or into the garden. I asked the previous owners about it and they said they seldom, if ever, saw them. The husband, who had originally come to the area many years previously from Orkney, did tell me a heartrending story of wildcat kittens that he found during the 1970s in a craggy area of our highest field beneath a gnarly ancient rowan tree, a tree I adore that has a hollow middle and still clings on with what seems a precarious toehold. He killed the three kittens, bashing them on the head with a road-mender's ditching spade, as they were then viewed as pests. It was a common story during that era – and now due to humans, and to hybridisation with the domestic moggy, the wildcat has almost completely disappeared from Scotland, and we are desperately trying to save it before it too is added to the growing list of extinctions. We are indeed fortunate that grey and red squirrels do not interbreed, so that the genes of the red squirrel, unlike those of the wildcat, remain intact.

A farm road borders the property and my goal was to plant a thick hedge for shelter and privacy, as well as to provide a valuable natural corridor and a food source for wildlife. Freddy and I also planted hedges around the garden using a mixture of beech, hazel, hawthorn, dog rose, holly and field maple. I didn't use guards and the high vole population revelled in our activity and gnawed through the beech plants in particular. I had to replace many on numerous occasions. Eventually, after a slow start and endless weeding, the hedges took off.

Soon after we moved to the farm there was a catastrophic outbreak of the dreaded livestock disease, foot and mouth. It had devastating effects on farmers and the economy, including tourism. Bans were put in place to stop people wandering in the countryside for fear of spreading it further, and everyone, including us, had a footbath of disinfectant at their farm gateways and notices telling people to keep out. One morning I had a call from a friend at Scottish Woodlands who knew of my passion for trees. 'We are currently unable to get out to plant thousands of native broadleaves, and we will have to burn them, do you want any, as I have a huge stock of bare-rooted plants here that I could bring you?'

When he appeared the next day with sacks and sacks of tiny whip-sized trees, it coincided with a heat wave – June is not the best time of year for planting anyway – but I had no choice. It was an incredible opportunity not to be missed. A heavily overgrazed area with a small burn running through it had been fenced off from the sheep. This was the spot I had chosen for my new wood. I spent the next week with Kim, my loyal Border collie, sweltering in the heat as if I were digging for victory. The tiny trees had first had their roots soaked in the burn, to give them a better chance.

Kim, a dog that I rescued, had been badly beaten by her owner, because she apparently would not work sheep. He had told me she was untrainable and disobedient. I had found quite the opposite. For twelve years she was never far from my side, she worked sheep beautifully and to date was the most obedient Border collie I have ever had. She has had no equal. Even now that she is long gone, a lump comes to my throat when I think of her – dogs do that to you. During long days of tree planting, she too dug busily in search of voles. And I was even able to utilise some of the holes she made to house some of my trees. It was hard but satisfying work. The heat continued and I wondered if all our efforts would be in vain. It was impractical to be carting water to nurture hundreds of trees. They would have to take their chance. I feared many might not survive.

I had overlooked the problems associated with the high local deer population, but luckily by June both the red and roe deer that normally help with the gardening tend to stay on the ground high above the farm. It is in the latter part of the winter that they are at their most destructive, when they are desperate for a fresh green bite. This is when our garden is given a thorough pruning. One winter, a particularly bonnie roe buck that came daily to eat fallen crab apples and to prune the roses around the door out into the garden, left me a gift of one his beautiful antlers, right on the doorstep. I viewed this as a thank you for all the damage he had done, and christened him the head gardener. So in order to save my new trees, I had to take out what felt like a second mortgage and purchase special protective tree tubes and stakes. And that involved another marathon of hard graft, fixing them all and hammering in stakes to secure them.

Most trees love being in tubes, except perhaps birch and aspen, both of which grow tall and leggy and then having grown too fast, have a habit of falling over. Tree tubes, while unattractive, serve as protective mini-greenhouses. Once the trees were all safely in their protective covers, I was advised to weed-kill a circle around each tube to stop the intense competition from dominant grasses. Twenty years ago, I did just that, but I now regret it. Though it helped the trees establish quickly, heaven only knows what it did to the environment. Since then I have not had obnoxious poisons anywhere near.

The trees took off beyond my most optimistic expectations. The aspens raced away the quickest but then had setbacks and flopped over like fainting soldiers in a parade so that they needed extra support. Though the farm is very sheltered, the aspens also caught the prevailing winds that drove down the valley from the west, and seemed to have only a tenuous hold on the ground. Five years later, I embarked on a second planting phase, expanding the woodland to include numerous trees and shrubs specifically for their wildlife benefits, including wild crab apple, hornbeam, guelder rose, sweet chestnut, lantana, whitebeam and Scots pine. I never burn anything, and tree prunings are made into habitat piles for myriad insects and birds, safe nesting and roosting places, as well as refuges for mammals. The local wren and robin population seem very drawn to these havens and there are always nests in their midst. We called the new wood Kim's Wood, and when she reached the end of life far too soon at the age of twelve, she was buried in amongst the trees she had helped me to plant. I regularly walk past her grave and will always remember the important role she played in helping me to establish what has since become a magical little sylvan oasis. I still miss her dreadfully.

Freddy hired a mini digger so that he could make two small ponds in the woodland for dragonflies and amphibians. Sometimes a heron comes in during the frog fornication season and takes advantage of the buy-one-get-one-free bonanza of frogs in amplexus – the smaller males clinging atop the females bloated with eggs. Before the wood took off, and now that the area was relieved of intense grazing pressure, a dramatic swathe of orchids appeared, but as a lush canopy began to close out the light, they vanished. In their wake came a profusion of bluebells, wood anemones, wood sorrel and primroses, as well as foxgloves.

Flag iris that I planted along the flanks of the little burn running through the heart of the wood began to spread faster than a rash. And when the sun shines through the leafy canopy, celandines and dandelions illuminate the burnsides in exuberant brilliance.

Ten years later, I was fortunate to be able to acquire a further piece of land next to the farm. It's a rich, thorny tangle of brambles and twisted lichen-encrusted hawthorns, blackthorn and hazel thickets as well as impressive ancient wild cherry – referred to as gean in Scotland. It was under threat of development for holiday chalets, but thankfully permission was refused. However, I feared this threat might raise its ugly head again, and I had started to see squirrels in this area of the wood, feasting on hawthorn berries and hazelnuts. 'That's great that you have bought it,' commented a neighbour who mows their grass shorter than a billiard table, and uses gallons of weed killer too. 'You'll soon get it all tidied up.' Tidying was precisely what I was *not* going to do. I have planted a few more trees around the perimeter, mainly hornbeams, in the hope that one day, as well as the newly arrived nuthatches, we might be lucky enough to lure a passing hawfinch, but I have left the open grassland in the heart to encourage fritillaries – in particular the small pearl-bordered and the pearl-bordered – and other vital butterflies and moths. This is where Iomhair keeps his beehives. And in May, when the hawthorn resembles a fulsome bride in an exotic creamy lace dress, the contented murmur of thousands of happy pollinators fills us both with delight. We sit on the woodland margins in spring listening to the chatter of the burn and the voices of newly arrived migrants – chiffchaff, willow warbler, blackcap and later, the cuckoo, while watching as comma butterflies emerge from their hibernation sites inside a few of the last remaining tree tubes. Commas didn't live in this area before, but they seem to be spreading north, and recently we are seeing more and more. Distinctive with their scalloped wing edges and

rich orange, burnt umber and brown pattern, they are well camouflaged as they hibernate in amongst leaves too. They seem to
favour the tree tubes as we have witnessed their emergence from
them on several occasions on the first balmy days of spring.

Most years since we took over this tangled area of woodland,
a roe doe has given birth to twins here. Once I stumbled upon a
twin-set soon after they had been born, lying flat, tiny and dappled
with great mascara-fringed eyes that blinked in fear. They were in
amongst long grasses and a patch of zig-zag clover where I often
find one with four leaves. I backed away in a hurry, not wishing to
frighten them further. It's at this stage that dog-walkers sometimes
find them and pick them up in good intention when they should
be left well alone. The doe will soon return.

This wood is also a maze of badger tracks. These cumbersome
members of the Mustelidae family have created a network of well-
worn paths. Some cross over the small burn, which was once a
mill-lade that ran right through the village and powered a lint mill
and a carpet mill. On up the bank the tracks lead to a flattened
area of earth where the badgers have squeezed their ample figures
under the fence, leaving wisps of revealing grey hair often tinged
with bronze highlights. Then once they have emerged into the
permanent grassland of our fields they bumble around, ever vigilant and ready to vanish if disturbed. They are usually foraging for
worms, wasps and beetles while doing copious amounts of excavating. They use their incredibly powerful front paws with their
sharp claws to dig out meadow ants, often leaving a large gaping
crater-like hole in the centre. Poor ants; disturbed and consumed.
Sometimes parts of the garden have had a makeover during these
nocturnal forays. I don't mind, the garden belongs as much to the
badgers as it does to me. They find unseen wasps' and bees' nests
and leave trails of comb and stickiness around the area, having
first consumed grubs or honey. Badgers, like us, love honey. Their

numerous latrine sites around the surrounding woodland reveal the wide variation in their menu.

And so we have christened the scrubby area of our little farm Badger Wood. In spring at dawn when the woodland steams and dampness rising from the leaf litter and sunshine tempts the new growth ever upwards, bluebells and birdsong are a marriage of pure perfection. Arias of blackbird, wren and song thrush fill the glades, moments of epiphany. And then I see flashes of bright ginger racing through the treetops and think that by now, the squirrels have probably already got kits hidden away in a secret bower, and any day soon they will emerge.

Brambles – blackberries if you prefer – are a dominant feature of Badger Wood. They spread their lattice of snare-like tendrils, and soon white or pinky-white flowers start to lure bees. Many of us overlook the humble bramble, yet it is an unappreciated fast-growing plant that is vital for so many species, from the tiniest insects and pollinators to a wealth of moths and butterflies. And squirrels, foxes, pine martens and badgers as well as bees, blackbirds and dozens of other birds relish the rich succulent fruits – the brambles, at the end of summer. Butterflies and wasps seem drunk on the fermented juices. Round the dense maze of bushes, I find flattened areas where the badgers have been coming in, ignoring the sharp thorns in order to push their bulky bodies through to reach the fruit. And the circle of life continues, for all the insects associated with this valuable plant in turn mean that a bramble patch is a motorway service station for small birds, and in particular is much-loved by members of the warbler family. Soon the bramble patch will host nests and there will be new life.

11

The Squirrels' Return

Some seven years after I moved into the new farm, I had the first excitement of a squirrel appearing on the birds' peanut feeder. To begin with there was just a single male – and though we do not usually give wild animals names, this one was nicknamed Baldrick as he was moulting badly and was certainly not a prime specimen. First he lost the top half of his little red coat, as usually happens during the spring moult, and then later the bottom half. Baldrick caused us concern as he was always scratching. Watching him through binoculars we could see that the large bald areas were bleeding and sore. His visits were erratic over the next three years, but with his very distinctively marked two-tone tail we continued to be able to recognise him. He would then disappear for weeks on end. Though there were clear signs of other squirrels in the area and we had plenty of sightings, few others actually came to the garden until the spring of 2012, when we had an epic squirrel year. It began with the appearance of another male, who had only one eye – this made him very easy to identify. Then two more squirrels appeared and began to use the special nut boxes we had put up for them. We bought a bag of hazelnuts in the shell – these were greedily consumed, though later in the year when I was planting bulbs and digging the vegetable patch, I discovered a great many of them uneaten all over the place and left them in case the planters remembered where they had cached them.

By late spring we could recognise four different animals, though the one-eyed boy had vanished. Over that summer three very tufty youngsters made an appearance too. We noticed that no two squirrels ever shared the nut boxes or feeders at the same time, as one always drove a new arrival away. There were constant arrivals and departures and when Iomhair put up a third box on the aviary, this too was used every day. We feed the squirrels a mixture of peanuts, hazelnuts and pine nuts – all of which cost us a fortune.

A fourth, much smaller, baby appeared later that summer, clearly from another litter. Its big feet seemed wildly out of proportion with its little body. It was quite hard to work out who went with whom, though it was possible to see signs of lactation in two adult females.

Scrub is a word that is often used dismissively – 'it's just scrub' is an excuse I often hear when I complain to people about their indiscriminate clearing and felling of important habitat both in their gardens and the surrounding area in spring and summer. It is usually also accompanied by the thoughtless comment 'the wildlife will go elsewhere'. And I always want to know, 'but where?' The farm is surrounded by scrubby woodland that is highly valuable for red squirrels and so much else besides. Largely consisting of hawthorn, blackthorn and large swathes of hazel, as well as a few beautiful mature pines, oaks and ash trees, it is good habitat for many species. Sadly, the ash is badly afflicted with the horrible ash dieback disease (*Hymenoscyphus fraxineus*), a fungal disease that causes leaf loss and crown dieback, and many are looking very sick indeed. As Dutch elm disease wiped out the elm, it seems that the same fate faces the glorious ash tree; its loss will leave a gaping void in the British countryside.

I continue to fill the garden with plants to attract wildlife and have amassed a collection of various crab apples planted in the hen run and small orchard. Blackbirds, redwings and fieldfares feast on the little brightly coloured fruits, particularly during protracted spells of hard weather, and the roe deer like them too. Due to the deer I have long since given up trying to garden properly and no longer buy exotic plants as they are quickly annihilated by these voracious cervine horticulturalists – roses are devoured with gusto so I have planted the wilder rambling types that actually seem to thrive with a little unselective deer pruning.

That first notable squirrel year was a blissful time for us. It was as if we were being given a little reward for all the planting that had taken place. Then one day in early autumn no squirrels came. Not even one. From days dominated by on average three peak visits – early morning, early afternoon and teatime, all the squirrels appeared to have vanished like snow off a dyke. It happened so quickly. We were very concerned. I kept telling myself that there was a glut of autumnal food out there; the wild hazel bushes were laden with fat nuts, and there were plenty of berries about too. The garden seemed like a desert without the captivating squirrel activity. We felt bereft. Friends who also have squirrels in their gardens rang to ask for advice. Why had their squirrels vanished like this? They had simply gone. This was not a gradual lessening of visits but more of a mass exodus. Worrying indeed. I reassured them that it was perfectly natural behaviour and that their visitors had probably found better wild food that year. But there were moments when it was hard not to think the worst. Had something sinister happened?

The nuts in the boxes went mouldy and had to be thrown out, and we continued to fret that the squirrels had either met some grim end or succumbed to disease – though thankfully currently there are no greys in our area, there are plenty of other diseases that might have killed them. Like parents that worry incessantly about their children, garden owners all around the country become so involved and attached to these delightful visitors that their loss, even for a few months, always leads to concern. Three months passed with fleeting sightings of squirrels in the surrounding woodland, but there were still none in the garden. Then one day, just as they had vanished, four returned. One-eye was there first, though he had been absent all summer, and then two females appeared together with another male. Even though most years the same thing happens, it never stops us from being fearful that one day our squirrels may not return. However, squirrels follow the food source; wild is always preferable, so when it's good elsewhere, off they go.

In the winter most of our squirrels grow spectacular ear tufts, though on occasions some have none at all. All of them have distinctive colouration and features and I try to keep photographic records to build up a detailed ID picture. It's impossible to count squirrels accurately, but the resident population appears to be fairly stable. Inevitably there are good years and bad, as is the case with all wildlife breeding cycles. And it's important to remember too that the weather will always play a part in the success and fecundity of every animal and bird.

We feed our badgers every night and sometimes pine martens visit us too. Both adore peanuts: the badgers perhaps surprisingly eat theirs very daintily but the pine martens don't have such good table manners. They too are erratic visitors, but sadly, pine martens, like badgers, are not always popular and it is likely, though you can never be sure, that some meet a premature end despite being legally protected. It's not unusual to have pine martens and squirrels in the garden at the same time, and though they are both aware of each other's presence, other than that the squirrels feed a little more warily, they seem unperturbed.

However, one night when I had two young squirrels in the aviaries awaiting release, a dreadful racket woke them both from their slumbers and the scene that was played out on a large aspen tree adjacent to their enclosure visibly agitated them.

It was a summer evening and we had finished dinner. I was heading out to shut the hens up when I heard loud yattering, and shrill, peevish screaming coming from the far side of the house. It was like neither bird nor mammal, but I reckoned it had to be a pine marten. Their vocalisations can be almost otherworldly, and some of the sounds they emit can be peculiar. A large male had been coming into the garden most nights for a month or so. He was always shy and vigilant but often stayed for some time until he had eaten all the peanuts in the feeder. Then he merged swiftly down into the beech hedging below, and was gone.

Creeping around the side of the house, I saw that there were two pine martens right up at the very top of a tall aspen, now the loftiest tree in the garden. Heads down and large hind legs fully extended, one was below the other, and both were clearly distressed. They looked like two flat fur rugs stretched out with legs splayed inelegantly. Below that, on the nut box feeder another pine marten was taking no notice whatsoever of the cabaret going on above and was greedily gorging on the nuts.

I crept carefully back to the door and Iomhair and I stood at the kitchen window and watched as an entertaining scene played out. Looking through the binoculars in the gathering gloaming, we could see it was a mother with her two almost fully-grown kits. They already seemed larger than her. She was oblivious to the fact that they were none too confident in their ability to descend from a height of around 25 metres. She was clearly at that parental stage where teenagers often give you a headache. In the adjacent aviary the squirrels raced around chattering their displeasure, having been rudely awoken by the loud protestations. Eventually, after a great deal more fuss, and a clumsy yowling retreat from the precariously wavering high tops, the young martens, their large paws splayed out and claws extended, tentatively made a slow and painful descent down to their mother. Once safely down they clearly forgot their ordeal and proceeded to muscle in, trying to get into the feeder at the same time, pushing and shoving her out of the way. The saga continued for a little while and then they all vanished, swallowed up into the shadows of encroaching night. For three evenings we were treated to a similar spectacle. It seemed they were not learning quickly and had not yet mastered the finer arts of abseiling.

Jays too cause us plenty of entertainment, and in early summer take over the garden feeders, consuming an astonishing amount of fat balls. Their varied vocalisations and ability to mimic often make it appear as if there are buzzards and curlews around too. Some of them are very good at imitating starlings – birds we disappointingly never have in our garden. These colourful members of the crow family also bring their offspring to the feeders, and the garden is noisy with their extensively varied gossip. Jays are always watchful, shy and untrusting, and don't hang about long if disturbed.

Nuthatches are a relatively new addition to the bird life of Highland Perthshire and in recent years have been steadily moving further north. In the spring of 2020 they started to prospect some of the nest boxes we have put up, and we were convinced they were about to use them. However, though there was much investigation and tapping and mud daubing around the entrances to the boxes to make them exactly the size they wanted, they disappeared. They reappeared in the garden later in the summer. It seems likely that they will breed with us soon. The nuthatch is a bird that can travel down a tree almost as fast as a red squirrel. The clumsy young pine martens should get some lessons from this adept little bird that resembles a masked highwayman and cleverly wedges nuts and seeds into the crevices of mature trees so that is can literally hack them open with its woodpecker-like bill. Hence the origin of the name – nut hack – nuthatch.

12

Natural Conflicts

As well as an increasing number of squirrels, red and roe deer, badgers and pine martens, many smaller mammals also come into our garden. Sadly, the few water voles that were once on the burn that runs through the property have vanished, but we have water shrews, bank voles, field voles and wood mice, and a large roost of pipistrelle bats, as well as a smaller one of brown long-eared bats. Hedgehogs sadly are almost never seen now, but the main reasons for that will be explained shortly.

The pine marten, like the red squirrel, was almost wiped out in Scotland, but over the past twenty years I have been increasingly seeing them in other parts of Scotland too. Each experience, be it in a wild, remote and windswept part of our diverse country or in the garden, fills me with a frisson of excitement and joy. Twisted scats are becoming more prominent on my walks, a sign that pine martens are beginning to emerge from the bigoted beliefs of the past.

The arboreal pine marten is a glorious member of the diverse Mustelidae family that includes the badger, stoat, weasel, polecat and otter, and like the rest of the group, it is adaptable and has many extraordinary traits. The Gaels called it *taghan* and in Scotland it was also frequently called *mertrick*.

In 1888, in his book *Wild Sports and Natural History of the Highlands of Scotland*, naturalist and sportsman Charles St John wrote:

> The marten cat is accused by the shepherds of destroying a great many sheep. His manner of attack is said to be by seizing the unfortunate sheep by the nose, which he eats away, till the animal is either destroyed on the spot or dies a lingering death. I have been repeatedly told this by the different Highland shepherds and others, and believe it to be a true accusation.

Supposed eyewitnesses backed up the theory that the hapless marten was indeed a livestock killer. Referred to, and treated as, the 'scourge of the glen', the marten is another example of an animal damned by Victorian dogma: sheep-slayer, a beast with a poisonous barb attached to its bushy tail, a filcher of poultry and eggs. The marten is guilty of only that last crime but people had vivid imaginations.

Nature-loving children often have a fascination with dead things, and in this respect I was no different. The deceased, providing they are not too far gone, can be closely examined, and much can be learned by simply looking and examining, The first pine marten I ever saw when I was a child was sadly a dead one, but it left its mark, and though it would be many years before I saw a living marten, I won't forget that first sighting. It was a beautiful dark chocolate-coloured animal brought into the public bar of my parents' hotel in Kilchoan, in Ardnamurchan. It had been caught in a snare but the fact that people were unable to confirm the animal's identity illustrates that at the end of the 1960s, even the Atlantic

Squirrels have long, elasticated bodies that enable them to leap through the trees effortlessly. (© Neil McIntyre)

The antlers shed by deer every year provide a valuable source of calcium for squirrels, who gnaw on them with relish.

Pipkin was one of the tamest hand-reared squirrels I have ever encountered. It is hoped that she will eventually be able to return to the wild.

This litter of three kits is ready to be moved outside to a large enclosure where human contact is minimised to give them the best chance of survival in the wild.

When I first moved to my little farm in Highland Perthshire in 1999, it had been heavily overgrazed and the land was bare and impoverished.

Twenty-one years on, following extensive planting of trees and hedges, the wildlife has returned.

The collective noun for squirrels is a 'scurry', but seeing these kits entwined around one another, a 'knot' would also be appropriate.

These were the youngest kits I have ever received – approximately four days old. Eventually, they were all successfully released.

With some of my charges. This litter was eventually returned to the wild.

Garden visitors. The glorious comma butterfly is increasingly moving further north and was seen more frequently in my garden in the summer of 2020.

Pine martens visit the garden on occasions and take advantage of the squirrels' nut box-feeders.

Badgers come into the garden every night and are very partial to peanuts left out for them. Sometimes they rearrange the flowerbeds, but the garden is as much theirs as mine.

The farm and Kim's Wood, looking west towards Loch Tay.

Numerous trees have been planted for wildlife. This crab apple provides food for pollinators and in autumn fruits for birds, squirrels and roe deer.

In May, the spectacular bluebell woodland above the farm comes alive with colour.

Tawny owls nest in the old bothy chimney adjacent to the house. There is always great excitement when the owlets first emerge from the maternity suite.

Having had a lucky escape following the collapse of their nest site in a derelict farmhouse, these owlets were successfully hand-reared. Once fully fledged, they were returned to the farm where generations of barn owls have thrived due to the unimproved, vole-rich pastureland.

Some years I receive numerous tawny owlets to hand-rear. Here you can clearly see the different colour phases – grey and rufous.

Despite the fact that owls are far from clever and have only very small brains, this one sensibly took shelter in a nest box during a downpour.

My son Freddy has always shown a great interest in wildlife and helped rear this tiny hedgehoglet.

An ailing hedgehog recovers well with a little spoonful of honey to tempt her waning appetite.

In autumn, hedgehogs that are too small to survive hibernation may need to be taken into captivity for the winter.

Cloudy, a red deer calf aged around a week old, was found in trouble close to the farm. Deer calves will not be released but instead live on the farm.

Iomhair adapted a baby harness to help Ruby, a deer calf brought in with tick-borne disease and paralysis.

Red deer really enjoy water. Ruby relishes buckets of water or a hose-down in spring and summer.

Squirrels are incredibly inquisitive and like to investigate everything, often first testing objects with their teeth.

Squirrels of all ages spend much time sleeping in between periods of frenetic activity.

Unimproved grasslands are an increasing rarity but provide valuable habitat for a wealth of species. The glorious emperor moth is sometimes seen during the daytime in summer on the farm.

A brown hare leveret amid a sea of wild-flowers in the traditional meadow adjacent to the farm.

The beautiful pearl-bordered fritillary is becoming a rare sight.

Left. These two young kits newly arrived from the SSPCA will first spend a few weeks in an aviary to adjust to their surroundings prior to release in the garden.

Below. As soon as young squirrels start investigating the china, it's time for them to move outside to the aviary.

Fleece beanie hats make ideal drey substitutes.

Using a syringe rather than a bottle with a teat means that milk flow can be precise and there is less risk of a greedy youngster inhaling the liquid.

When in captivity, squirrels will eat a surprising range of food. Sprouts often prove popular, but not every squirrel likes them.

Ruby surveys the scene high above Kim's Wood, where the new trees have grown exceedingly fast.

Parasites are a feature of all squirrels, and fleas and mites as well as ticks can cause a great deal of scratching.

A particularly enchanting little squirrel, Pipkin, soon after her arrival.

Most baby squirrels grow luxuriant ear tufts.

With Cloudy, writing *A Scurry of Squirrels* during the summer of 2020.

oak woods of the western seaboard were devoid of pine martens. We stood looking at this astonishing creature amid conjecture – was it a ferret, a polecat, a mink? No one was sure. Someone suggested a marten cat. This lack of knowledge proves how desperate things had become for the pine marten. It had become so rare it was almost lost from our memory. Surely these oak woodlands would previously have harboured many pine martens that were historically Britain's second most common carnivore?

Like the red squirrel, the catastrophic loss of native woodland due to fire, axe and overgrazing, made survival impossible for the pine marten. And like the squirrel, its luxuriant, exotic pelt also added to its downfall. In 1906 naturalist J.A. Harvie-Brown claimed that it was even rarer than the wildcat. The few remaining animals fled to the remotest corners of Scotland, where they adapted their superb tree-climbing skills to scaling rocky outcrops where they sought dens high in inaccessible crags. Here they miraculously clung on by their pale-coloured, semi-retractable claws and were seldom seen. Squirrels on the other hand were unable to adapt and died out completely in many of their old haunts.

In 1988 an amendment to the Wildlife and Countryside Act of 1981 brought protection for the pine marten, and as mentioned earlier, the red squirrel. Around a decade before, the mass planting of fast-growing commercial conifers also inadvertently aided a population recovery among the pine martens. They are omnivorous, with a catholic diet. And despite what some people will tell you, they don't survive solely on red squirrels, rowan berries and hens but relish earthworms and field voles, as well as numerous other foods, and the long grasses of the young forest fringes brought rich pickings. As the monoculture tree canopy closed in, it offered safe, secret places to hide.

Today the majority of us no longer want to view the pine marten that previously swung husk-like from the gamekeeper's gibbet through the barrel of a gun. Instead, the *taghan* has become a figure-head for a growing wildlife tourism industry. From a once-detested varmint that polarised people's perceptions, to an ambassador for an ecologically richer Scotland, the pine marten offers a glimpse into what might be possible with a change in our attitudes.

As well as resurgence in Scotland, martens are beginning to leach into Cumbria and Northumberland. However, it is the exemplary work of the Vincent Wildlife Trust (VWT) and its partners that is helping the beleaguered population further south in Wales and England.

Social acceptance is a critical factor in the complex process of re-establishing any lost or endangered species. For such projects to succeed, public opinion must always be a significant consideration. Following two decades of detailed consultation and surveys, the VWT found that 90.9 per cent of respondents wholly supported the idea of helping restore the pine marten population in mid-Wales. Scottish Natural Heritage (SNH), now renamed NatureScot, granted licences for some animals to be taken from areas of high population density, and between 2015 and 2017 a total of fifty-one martens were successfully translocated to mid-Wales. The Welsh releases took place in secret forestry locations, and follow-up radio tracking and camera trapping has revealed that every year since, breeding has occurred and the martens are flourishing. In September 2019 the VWT reintroduced eighteen martens to Gloucestershire's Forest of Dean. The hope is that this latest group of animals will eventually link up with the Welsh population. As I write in 2020, there is now evidence that these martens have already bred and that the reintroduction has proved successful and shown that the habitat is ideal for them. It also shows what we can achieve collectively to restore nature to its rightful place, and brings a glimmer of positivity at a time when we have never needed it more.

Those who clutch at anti-pine-marten straws are quick to suggest that their population recovery will be detrimental to our struggling red squirrels. However, following detailed research led by Emma Sheehy of Aberdeen University both in Scotland and Ireland, the pine marten's role in red squirrel conservation has proved fascinating as well as surprising.

We know the detrimental effect that the introduction of the non-native North American grey squirrel has had on red squirrel numbers. But where pine martens are regaining a foothold, the grey squirrel population is tumbling. Conversely, and perhaps surprisingly, reds are thriving. Not only is the red squirrel swifter, lighter and more agile than the grey, so it can better escape predation by reaching the thinnest branches, it also co-evolved with the marten.

One of the most exciting results of this research is proof that the presence of native predators can help naturally with the removal of invasive species. Had the pine marten population been buoyant when our misguided forebears brought greys to the British Isles, would we still be faced with the problem of their insidious spread? This developing research also brings to the fore a basic ecological principle: that the removal of a predator from an ecosystem often results in catastrophic consequences. For me, the memory of that first deceased marten lingers, but half a century on things are indeed changing for the marten cat just as they are for the red squirrel.

I often reflect back to one particularly atmospheric pine marten encounter on the wild shores of lonely Loch Maree in Wester Ross. The eerie, melancholy calls of black-throated diver echo beneath the bastion of brooding Slioch. On the mountain's lower slopes, where wild goats graze, the last remnants of ancient contorted Scots pines bring memories of a wooded past. On a moss-upholstered boulder close by, a pine marten dozes. It is cat-sized, curled into a neat circle – ever vigilant. If I move, it will fade into the landscape as if it was never there. I lie hidden in russet-coloured bracken beneath the bank. Through binoculars, that lustrous pelt once so eagerly sought, glistens with raindrops as drizzle falls fizzing onto the gunmetal grey loch. The creamy-yellow cravat, with its unique brown patches, makes it possible to identify individuals. There is movement. It rises and stretches out its long, athletic body and gallops up into the crags above. Sounds of yattering, growling and hissing indicate a dispute – two martens emerge facing one another in a standoff. Rain is running down my neck; the loch has vanished.

Loch Maree and the last vestiges of Caledonian pine forest now at the very end of its life, is a place that would once have had a healthy population of red squirrels too. This previously rich woodland was decimated during the seventeenth century, when pine was felled for ships' masts and the surrounding oak woods laid bare so that the timber could be used in the leather and tanning industry, and also to fuel the blast furnaces for a thriving iron industry. But when the habitat recovers the squirrels will come back, thanks to some incredible inspirational work to translocate them that I shall explore later.

*

Our attitudes towards predators are often deep-seated and misguided. Some stem from cultural beliefs that can be traced a long way back. Some are entirely due to fear and lack of knowledge. After just a few decades of recovery, there are those who consider that already there are 'too many' pine martens. In most cases 'too many' translates as more than there were ten years ago. Martens, like red squirrels, were indeed once abundant across the British Isles and their return stands out as a beacon of hope in a sea of ecological despair.

If we think back to the situation with the red squirrel, it too was mercilessly persecuted and was teetering on the sheer cliff-edge of extinction, having been culled relentlessly because there were deemed to be 'too many'. Both the marten and the red squirrel are priceless pieces in our sylvan jigsaw, and in a healthy ecosystem they co-exist perfectly.

Altering our attitudes towards predators while accepting the essential role they play is a critical component of understanding that habitat restoration – rewilding, and letting nature into our lives – will benefit all life. The pine marten's return and the change in our attitudes demonstrate another small step forward on that exciting journey, following where the adored red squirrel has already started to lead us. But understanding wild things and the way they interact is vital. In the case of the relationship between pine martens and red squirrels, if we are someone who does not like the former, then blaming it for killing the latter is easy. Instead of doing that, we have to look closely at the role we – as the top predator – are playing in these complex relationships.

On that note, there is another worrying trend involving the blame game. Hedgehogs are in serious trouble. Since the 1950s, when they numbered many millions, their population is estimated to have crashed drastically, and as usually happens, we search for something else to blame instead of ourselves – in this case badgers. Hedgehogs, as their name suggests, rely heavily on species-rich hedgerows and connected native woodland. Intensive farming, forestry and development, plus an increased road network and chemical sprays have led to a paucity of suitable habitats for an animal that does not live underground. The badger, which lives a largely subterranean life, is in most cases better protected (if we disregard the horrendous issue of badger-baiting). Like the hedgehog, it too is an omnivore and competes for a similar diet. In an invertebrate-depleted environment where there is not enough food to go around, it is obvious that the larger, more powerful animal will come out on top. With little or no cover for hedgehogs to move around in safely, they are increasingly becoming easy prey for an opportunist badger.

When two species compete for the same food source and there is not enough, then one may prey upon the other. This is known as an asymmetric intraguild relationship and is common in nature.

I occasionally find the dry, hollow skins of unfortunate hedgehogs that have served as a badger snack. Nature is harsh, but it's ecological madness to contemplate wiping out badgers in order to conserve hedgehogs. Don't forget the fact that as I said earlier, the dear little red squirrel sometimes takes songbird chicks and eggs from unguarded nests, and that a woodpecker drills out tit boxes to devour the young within. It's normal behaviour. Hedgehogs may eat baby birds and their eggs too.

Most mornings, heartrending petitions fill my burgeoning inbox. Nearly all relate to the mounting global crises in the natural world: climate change, habitat destruction, sprawling development, pesticides, chemicals, poaching, whaling and appalling animal cruelty. It's an interminable list that leaves me overwhelmed with sadness. What have we done? I feel utterly hopeless and helpless, but sign anyway. If we don't put the natural environment before everything else, nothing will be right. Supporting petitions, though, is not enough. We have to do far more than sit in a nice warm house basking in artificial light at our computers, living a lifestyle that adds to the problems as we sign up. I am as guilty as anyone.

In the rural environment, hedgehog numbers have fallen by around 50 to 75 per cent, while in urban areas the drop is around 30 per cent. If an omnivorous generalist such as the hedgehog is struggling, then what does this mean for everything else? I have been closely following the valuable work of Hugh Warwick

– Hedgehog Hugh – for many years. He is the country's finest hedgehog ambassador and one of his petitions was a brilliant campaign to lobby the government to bring in a law applicable to all developers and building companies. It's a simple plan to ensure that in the massive rash of new housing schemes countrywide, all adjoining fences between properties will have a 13cm square hedgehog-sized hole built in. This would allow hedgehogs to pass from one garden to the next on their nightly foraging trips. Hedgehogs may have a home range of up to 30 hectares and may travel over 2 kilometres a night; they need to be able to move freely and safely to find enough food, as well as a mate. Increasingly, they are restricted because there are barriers everywhere. It would cost less than 50p per fence for each building company to implement this vital lifeline.

When I shared Hugh's petition on social media, I was shocked by some of the vitriolic responses from ill-informed people. Social media can be treacherous – heated debate achieves nothing. I resisted the temptation to fire back an explanation to dispel the bigoted viewpoints. A polite version of this aggression suggested that, rather, we should remove every last badger from the British countryside. It was suggested that badgers were 'wicked, evil' creatures – vermin that caused nothing but devastation. Indeed, it was, they said, solely due to the presence of these 'vile' mustelids that hedgehog numbers had plummeted. Many of the comments suggested I start a petition to rid the entire country of this vicious devil instead. It's not the first time this argument has arisen. It has been simmering for decades.

Badgers are an integral part of our ecosystems just as much as hedgehogs and red squirrels. Most naturalists would agree that we cannot choose one native species over another and simply wipe it out because it does not fit in with current human notions of 'good' animals versus 'bad' ones. I want to set the record straight: if our ecosystem were healthy, then wildlife would find its own natural balance. When it comes to environmental matters, seldom do we take a long hard look at our own actions. What we are doing, and have done, is almost always at the root of these natural conflicts.

13

Ruby

June 2010

It's a moody June afternoon, and I have been dealing with a host of unexpected wildlife-related problems, prevalent during the breeding season. A man rings and regales me with a convoluted tale of a red foal.

'A foal?' I ask him.

'Yes, I found her upside-down, lying in a ditch. I took her to the vet for treatment yesterday and we have had her since but I really don't think we can look after her long term.'

I feel confused – why would he be looking after a foal that would surely belong to someone? 'A red foal?' I query again.

'Sorry,' he adds, 'I don't mean a foal, I mean a red deer calf.'

Later, a couple appear with the calf in the back of their car. She is encased in a thick high-vis jacket and lying on the back seat. She looks like a cervine equivalent of an Egyptian mummy smothered tightly by the jacket. She is sick. Her head is poking out of the folds of the jacket and only the whites of her eyes are showing. Her head is lying back on her body in that ghastly position that I have come to dread with any animals – it often indicates that death may shortly follow. The calf is hot and distressed and the heavy jacket is making her even more uneasy. It's another case of no hope but I don't want to appear negative, as they tell me that she is already far better than the day she was found. 'She's actually doing very well but we really cannot keep her.'

I had made a pen in readiness, and my immediate reaction is that she needs to be left in peace to cool down and settle. Being less than a fortnight old, she is vulnerable. As I leave her in the straw, I turn to have one last look. Two huge brown eyes roll uncontrollably, revealing their whites. It really does look hopeless.

We sit around the kitchen table as the charming couple tell me more of the calf's story. He is a handyman on a large local estate and he had been out fencing when he found her lying in the ditch. He picked her up and took her to the local vet, who diagnosed a massive infestation of ticks as the cause of her immobility. He gave her strong antibiotics and steroids, and electrolytes to replace fluid loss. 'She was so badly dehydrated when I found her,' the man explains. And she still looks in dire need of fluids, I think to myself. After they leave I note that her skin when pinched remains in a tent-like shape, indicating that she does need fluid urgently. I mix a weak solution of glucose and milk powder, and endeavour to get her to take it. But she won't suck or swallow and seems paralysed. Eventually, I inject fluids under the skin, and within a few hours she has perked up considerably. Once she has recovered from the shock of the journey, there is an aura surrounding her that I have never witnessed in all the years I have been working with injured and orphaned animals. She has a will, and brightness. It's extraordinary to stand and watch her large ears moving expressively and the beautiful eyes no longer showing their whites, staring nervously back up at me.

During my childhood in Ardnamurchan, while driving home in the gloaming red deer dominated our journeys as they crossed in front of the car, shining eyes and large dark shapes that loomed bigger in the headlights. My father adored deer. He was right when he told me that their natural history is one of the most captivating wildlife subjects. Deer have always intrigued me, too. I never cease to be amazed by the way in which stags shed their antlers annually; new antler is the fastest-growing mammalian tissue on Earth. And the time of the red deer rut in October has always been one of the highlights in my wildlife calendar. There is much to learn about our largest land mammal, and though currently numbers in Scotland are very high, and there is little doubt that the pressures of overgrazing are leading to severe land degradation, they are a valuable and integral part of our fauna – beautiful, graceful animals that also provide a healthy source of unadulterated meat.

Towards the end of May and throughout the month of June, their perfect dappled calves are born. From an early age I loved to try to find the season's first calf and then to watch as more and more appeared and they skipped beside the attentive hinds, or suckled quietly. I loved to hear the soft contact calls across the bracken-clad slopes and to watch as a group of hinds and calves ventured higher on the wind-bruised ridges to avoid the aggravations of the increasing number of flies. You can learn so much in quiet contemplation if you too can tolerate the inevitable biters that accompany the Scottish summer. I knew some of the likely places to find very new calves, the places where the hinds broke away from the rest of the herd as their time to give birth approached. They often chose a secret spot in amongst rushes. A particular punchbowl on the flanks of Ben Hiant, the hill that dominates the village of Kilchoan where we lived, was a likely place to find a camouflaged calf, hours old, curled up sleeping, lying in wait for its mother's return, its pale spots a perfect imitation of the landscape of rippling bog cotton. I secretly hoped that someone would find an orphan calf and give it to me to bottle rear, but though pet lambs were always a dominant feature of my childhood, no red deer calf was forthcoming. Though I found many, nothing would have made me steal one away. We are indeed a poor substitute for an animal's own mother. My own mother had always made sure I was well aware of this fact. 'Never take birds' eggs, just look, and don't pick up baby birds unless you know they are in severe danger,' she told me. Though over the years I have since hand-reared roe fawns – and they are far from easy, probably one of the worst wildlife patients for succumbing to stress – for much of my adult life red deer calves have been elusive. Now one was in my care, but her life hung on the thinnest of threads.

The calf's bright aura and will to live seem tangible. A neighbour, Margaret, spends hours too with the calf. She is an excellent, patient nurse when I am not able to be there. Much of a creature's recovery is dependent on nursing. I have had some miracles where animals have survived terrible ailments or injuries. Often there is little doubt that recovery is also down to dedicated nursing care.

As someone who advocates the planting of more trees, who understands the effect that too many deer on our hills has had – no natural regeneration, and millions of hectares of bare, impoverished ground – sceptics might ask, why bother with this particular calf when there are so many deer in Scotland, and they are hardly endangered? Why did I? I bothered because I thought at that moment that the calf was one of the loveliest beings in creation. I have known many honed deerstalkers and keepers of the old school who have done the same, and who, despite the fact that the main aspect of their job is to cull deer, have kept one or two as pets for many years.

I had instantly decided that if she did recover, which seemed dubious, this red deer would be our responsibility for her lifetime. This was not a wildlife casualty that could be released. I also knew that red deer hinds are easily domesticated. On Ardnamurchan Estate I had spent many days out on the hill with Hugh McNally, the stalker on the estate when my family owned it, and brother of the great naturalist, author, photographer and stalker Lea McNally at Torridon. Hugh had arranged for me to meet Lea and to see Beauty, his pet hind. Beauty was also unusual because she once surprised her owner by giving birth to twins – something that seldom happens with red deer, unlike roe, which frequently have two fawns. I had also been enchanted by a story from another old stalker friend in Sutherland who had a tame hind that he took in the van with him on occasion. During the stalking season she would follow him out on the open hill when he took guests out. She would lie patiently waiting and chewing the cud while a shot was taken. After this, apparently totally unperturbed, the hind would then follow the party home again. Deer are far more intelligent and intriguing than many people give them credit for, and they demand our ultimate respect. I understand that there is a pressing need to lower their numbers but advocate that this should always, without exception, be carried out only under the jurisdiction of professionals, and the welfare of the animals must always be the priority. Shooting deer from helicopters, and killing hinds when they still have dependent calves at foot, is both abhorrent and of serious welfare concern. Hugh McNally told me that unless an animal is wounded, it should never be shot at when it is running, or in water – often the sea. He also told me that when it comes to hind shooting, you must first shoot the calf – the dependent – and never shoot the calf's mother first, in case the calf is left orphaned only to die a protracted death of starvation over the harsh winter months. Sadly, that does not always happen today and I often think that he would be appalled if he knew of some of the cruelty that takes place, the total lack of respect and professionalism. I feel very honoured to have spent time in the company of such people as Lea and Hugh McNally. These were true countrymen with vast knowledge, who also loved and understood deer.

While there were personal reasons for my desire to nurture the new arrival, both Iomhair and I wanted to see if we could beat the damage that had been done to the calf by the heavy infestation of ticks. There is little doubt that the tick is the most dangerous creature in the British countryside – minuscule yet with the power to severely disable and kill. As ticks are becoming more prevalent, it would be a major coup if we could pull the calf through this, and we might learn something in the process. More and more valuable farm livestock as well as domestic pets are lost to ticks every year, and they are a huge threat to us too. Lyme disease is on the increase and the more we can learn about ways to tackle the tick issue, the better. I have always been fascinated by parasites and their extraordinary life cycles, and that of the tick is a complex and dark world.

Next day I rise in trepidation. Lovely long donkey-like ears move expressively and shining dark-brown eyes peer up at me from deep straw. Deer calves and fawns even from an early age are very dependent on browse – but the freshly cut leaves we'd left out for her to eat remain untouched; she has not moved. But she is alive. That alone seems like a miracle. When I first saw the calf my immediate reaction had been that she should be put to sleep. Sometimes it's very important to be practical too.

Two vets come to see her at different times, and give her more strong antibiotics, steroids and vitamin injections. They too are very keen to see if we can cure her of the tick-induced paralysis. The repulsive, bulbous grey ticks that cover her have to be removed as soon as possible, since it seems that the toxins in tick saliva are the cause of the paralysis. Her legs remain powerless, the muscles useless, as is her sucking reflex. Should we really keep going? Getting any fluid into her is a struggle; Margaret and I in turn sit there for long periods, squeezing it gently down her throat. Despite this the calf remains remarkably alert and interested and does not have the worrying listless aura of an animal that is about to die. Normally we tube-feed ailing lambs or calves, but I am worried that if we try this with her, the stress might tip her over the edge. The calf loves company and is visibly distressed when we leave, emitting a soft contact sound that I have often heard when out on windless days on the hills. We move her about in the pen and keep her straw clean to ensure she is not lying in her own burning urine. But after a week, when she can still not get up, I am beginning to think that we really should let her go, for she has no coordination whatso-ever. Iomhair comes home one afternoon bearing a baby harness he has acquired at the local charity shop and cleverly adapts it to fit her. Now we are able to hold her body up so she has to put weight on her useless legs. Ragdoll-like, she wobbles and falls each time we try to get her to stand.

I had read on an American website about the perils of ticks, that once every last one is removed, if the creature has survived thus far, there might be the possibility of a recovery. Usually it is only one single tick causing the problem, though having such an infestation of bloodsuckers can also lead to anaemia. A Belgian vet working locally has never treated deer before and knows little of ticks. She is fascinated and spends a long time picking them off with me. We chat over the little bloated parasites as we pull them off and let them fall into a bucket. There are dozens and dozens. We had already applied a pour-on tick product, and though the horrible things are now dropping off and easier to remove, it takes us time. We lose count after 200. It's not a pleasant task, but we are both also filled with a morbid curiosity about the power and insidious nature of these tiny bloodsucking members of the spider family.

Though I remain in doubt, and am trying to keep sentiment out of this, we are all determined to win the battle. In the back of my head rumbles a depressing thought that it is all going to end in failure. Though I am trying not to, I have become besotted with this gorgeous animal with her extraordinary spirit. I am beginning to love her.

Then there is a breakthrough. The calf suddenly starts to suckle, and the bottle of milk vanishes amid frothy saliva and enthusiasm. And then she is looking to me for more. It is as if someone has pressed a switch and turned a light on. Her legs remain weak and useless, and seem to splay in all directions. We massage them to stimulate blood flow. We have been taking her into the field and helping her to stand, and have been propping her up between bales. Is there perhaps a tiny glimmer of hope? I decide that if she cannot get up unaided after a month, then we must let her go. It's the only sensible option. It hangs over the three of us, for the hard fact remains she cannot stand up unaided. Then after a fortnight I go down to her pen to find her upright, albeit wobbly, but by herself. This is the spark we need to continue.

I have resisted the temptation of a name till things look more positive, and now Ruby is officially christened. Progress is slow, but she is relishing four bottles of milk a day and visibly gaining strength. I put her outside for a spell each day, but she hates being left and makes a surprisingly noisy fuss. I am hanging out the washing when I hear a crash and turn round to see that Ruby has jumped through a gate and is standing quaking beside me. Our paralysed calf has jumped.

We put her in a dog harness and use it to guide her to walk carefully as then her legs work quite well, but it is difficult because she is so determined. On several heart-stopping occasions, she bounds off with the speed of a bullet and has some horrible falls. It's upsetting to witness her crash landings. Once again my worries resume and I fear that in the end we may still be defeated.

After weeks of heavy summer downpours and high humidity, the grass in our fields is long and lush. It does not make an easy place to run for there are thick tussocks to trip over and the ground is a snare of dense growth. We take her to the garden with its mown sward. She explodes forth for a few yards, and then turns and comes straight back to me, mewing in the familiar contact call she now makes each time I leave her sight. I have been working on another book – *Fauna Scotica: People and Animals in Scotland* – and it seems apt that much of it is written in the shed, where I make a makeshift study with Ruby by my side. She picks quietly at leaves and then lies close to me cudding rhythmically, her eyes closing, head lowered onto her spotted side. Peace.

Deer fencing is another problem. The hen run is secure, and to begin with she goes in there for the day. She is curious about the poultry and cheekily chases them around. Amid flying feathers and indignant squawks, our crippled calf is blossoming. The hens join in the chases, pecking at her coat as if to goad her into another game. I remain philosophical about the future, knowing that her weak legs are still far from right. We are, however, moving along in the right direction.

We want to deer-fence our fields because with a new path network running above us, and an increasing number of out-of-control dogs about, our sheep are vulnerable. Adding high fencing is not an issue. In fact, if I want to sleep soundly, it's paramount. Sadly, there appears to be less and less respect for livestock and farmers are increasingly witnessing dreadful problems of sheep worrying from dog attacks, with major losses in some areas. The family pet may seem benign and adorable, but every dog has the potential to become a sheep worrier, or worse still, killer. Fencing dogs out while also keeping stock in seems the only sensible solution. One night we take Ruby out and she takes flight so fast we are awestruck. She goes right through the middle of the sheep. They have already met through the fence but are most surprised, scattering in all directions. However, she is running almost perfectly, her sharp black hooves covering the ground elegantly as she skips lightly over, putting in the odd buck and high kick.

Ruby now lives with the sheep. She has a particular companion, a small pied Shetland ewe called Juniper that she is often with, and they are very affectionate with one another. Red deer love water. On hot days, she mews to me over the fence and relishes a hose down, and while I am cleaning out the water troughs she canters over to have buckets of water thrown over her by me. But she won't let anyone else do it. There appear to be no long-term effects from her life-threatening tick infestation. Having a red deer hind as a close companion is not only very special but it has added further impetus to my desire to continue learning more of the natural history of our beautiful red deer.

14

Squirrel Tales

All the squirrels that are brought to me are memorable, but some are far more so than others. Sadly, many of them die soon after arrival. If cats or dogs have caught them then often they stand little hope of recovery. And for something so small to come into contact with a car or a van, then it is lucky indeed if the subsequent injuries do not lead to death. It's heartbreaking when at first glance I am instantly aware that there is nothing I can do. However, as I look back through my wildlife diaries, many of the success stories are noteworthy. I have always been over-emotional, and there is no doubt too that reflecting back brings a lump to the throat. Squirrels do that to me, but then so do all animals.

Despite the long list of casualties that have died in my care, there have, thankfully, been many occasions when there is an amusing tale attached to my encounters with injured animals. When I first came to the farm I was fortunate to have wonderful neighbours, Tim and Pip. Two retired professionals who were wildly eccentric and full of fun, they had numerous children and grandchildren who lived in North America. Some of them were angst-ridden and troubled, battling with fluctuating hormones

and the accompanying unwanted pimples – few teenagers escape
this difficult time. At the height of a monosyllabic, sugary cereal-
eating phase, they would arrive, pale and bored with life in
general. But an extraordinary transformation would take place
during the time they spent with their grandparents in their school
holidays. Following a strict regime of den-building, picnics, swimming in rivers and lochs, hill-climbing, camping and reading, they
all visibly blossomed during their sojourns. Sometimes they came
to spend time with me to help with the chores around the farm.
All of them loved animals; it's joyous to be able to further fuel this
interest and also to steer children towards a lifelong connection
with wildlife and nature. Some have it from the outset – others
quickly learn the importance of a love and respect for the natural
world.

Many people, and in particular teenagers, miss the best part of
the day by lying abed for far too long; Tim and Pip would have
none of it. It simply wasn't part of the remit. Their disgruntled
grandchildren would sheepishly appear early in the morning
bearing a picnic, having been booted out. The grumpiness quickly
diminished; few activities are more levelling than being outside
in all weathers, and partaking of a little physical labour or exercise is a tremendous way to improve the mood. Helping with the
annual task of barrowing steaming dung up to the vegetable patch
in readiness for planting the tatties seemed to keep everyone on
track – back to basics and a connection to the earth.

One of their grandsons, aged around twelve at the time, asked,
'What exactly is this stuff, Polly?' as he shovelled more and more
of the black ordure, dark after years of maturation, into the barrow,
and we bent down to collect dozens of squirming worms from out
of its midst to feed a brood of orphan blackbirds that I had at the
time. I told him it was dung produced by the sheep, goats and
hens. There was a long silence accompanied by a look of horror.

He couldn't believe this was to be used to fertilise the ground to grow potatoes and other vegetables that eventually we would eat. After his return home to Canada, he sent me a wonderful letter: 'Thank you for having me, I had a great time and I loved all the animals even though I learned the terrible truth about dung and what you do with it.'

Three very girly, urbanised granddaughters came to stay one summer when there was a spell of extremely hot weather. We all threw our house windows open wide day and night, but still there was little respite. A stifling atmosphere continued to engulf us for several weeks. Our gardens wilted, the grass turned brown, and so did we.

I had recently received an injured squirrel. On arrival it had looked in a very poor state, lying flat, almost unconscious and scarcely breathing. I was not optimistic about its chances. However, the vet gave it some fluids under the skin, and then I put the weak little creature onto a hot water bottle covered with towels in a dark box. I was dubious whether it would survive the night.

At dawn I was astounded to see that it was still alive. Barely. Its breathing was loud and laboured, and watery blood trickled out of its nostrils. For ten days that squirrel lay on its bed and didn't move unless I changed the towels and hot water bottles. I injected it with sustenance several times a day. Few squirrels in this state survive this long.

On the eleventh day, it raised its head and shook itself. When I offered it fluids from a syringe it drank willingly and continued to take fluid all day. Then it began to show interest in the food that I was putting into the box. It had eaten nuts and nibbled at thin slivers of apple. It moved lopsidedly around the box. I could hardly believe it. After another week in the hospital box cage, it was still not very agile but was eating and drinking well by itself. It was time to put it outside into the aviary, where I hoped exercise would quickly rebuild its strength.

After another week it began to make furtive dashes for the door every time I appeared with food. I wanted to release it back into the pinewoods close to where it had been found, and had no intention of releasing it at home, where there were far too many predatory cats that would merely have added another hazard for the poor thing to deal with.

But it was not to be. Foolishly, I decided to give it a proper clean out and as I was removing some faded branches from the cage floor I tripped and knocked open the second door. A flash of orange like an exploding firework shot towards me, and using my head and back as a springboard, it bounced out and disappeared into the woods. My language was unrepeatable. I knew I had no chance of catching it again, but felt that at least it felt well enough to go.

Unbeknown to me, the three teenage granddaughters staying next door were upset to find that every night when they went to bed there were little torpedo-shaped droppings all over their lacy pillows. Everyone in the house, including Tim and Pip, was convinced these were rat droppings. Panic set in – something had to be done. And fast. Each night there was a thorough investigation of the bedding and close inspection revealed more and more of the same. The girls were not happy and became paranoid and hysterical about bedtime.

For a week, I saw no sign of my squirrel. Then I caught a glimpse as I was driving out one morning. It had a particularly dark tail and was easy to recognise. It was sitting on a different neighbour's woodpile, holding a cone in its little front paws feeding greedily. It looked enchanting.

The intense heatwave continued. Even at night it was relentless. I had gone to bed early one night when the telephone rang.

It was my neighbour Tim. 'Have you lost a squirrel?' he asked with a twinkle of humour in his voice. It was about 11.00 p.m. and as the telephone had awoken me from a very deep sleep, I was somewhat confused by the call. I hadn't a clue what he was talking about. 'Have you lost a squirrel?' he asked again. I took some time to come round and wake up properly. Apparently, the aforementioned teenage trio had just been completing their ablutions and were climbing into bed when suddenly a little cheeky face had appeared from behind the curtains. Their loud screams terrified the hapless squirrel and, completely crazed, it started to do loop the loop all around the bedroom, rushing up the pictures, springing off the beds, and belting round the top of a wardrobe. Later they described a hilarious scene of a squirrel going into orbit. And the story grew more and more legs.

Now the voice at the other end of the telephone was asking me if this squirrel belonged to me. Miraculously, despite the hysteria, Tim had managed to capture it by luring it into a cat basket with nuts. So the squirrel came home again with its tail between its legs to recuperate from this latest ordeal. After another week of dinner, bed and breakfast, we deemed it fit for release again. This time we took it far up into the pinewoods, where there was an abundant supply of natural food, in an environment safe from encounters with teenagers and cats. At least the mystery of the unknown droppings on the pillows was solved. And how many of us can say that we are sharing our beds with a squirrel?

*

March is a month of extremes. I often think of it as the harshest part of the winter, yet equally it can be as benign and cheery as the new lambs that appear in fields all around Highland Perthshire. A day beset by gales, sleet and hail usually means there will be calls about wildlife in trouble. Mild spells in February are conducive to mating, meaning that squirrels in our area often have litters by mid-March.

Our vet rang to tell me that the strong winds and torrential rain had dislodged a drey and in it were three tiny squirrels. Someone from Aberfeldy had found them in her garden and had sensibly left them for some time in the hope that the mother squirrel would return. She had waited for as long as she felt was safe given the bleakness of the day and the savage cold, and when no mother returned, she had taken the drey and its precious contents to the vet.

We were fortunate to have vets who had a keen interest in helping with wildlife, and during the time I have been on the farm here, they have been the source of some memorable patients. They were most concerned about the plight of the squirrels, and after ringing to check I could take them, sent them to me with a tin of kitten milk substitute.

When she arrived she was worried that they wouldn't survive. I looked inside the thick nest of moss and lichen, and felt concern too. The babies were so tiny, but not half as tiny as a litter I would receive in the future. These ones were very active and moving about more than I would have expected. And they didn't appear to have been chilled, despite the bleakness of the day. I quickly made a hot water bottle and transferred them into a suitable basket. As is always the case, the drey was moving with dozens of fleas, and so were the squirrels.

Firstly, I needed to get some fluid into them. I made a very weak solution of milk and added glucose, and tried to feed each with a syringe. They didn't want it and refused to swallow, so I had to tease the fluid into their tiny mouths. It's vital not to give them too much to begin with as they can easily choke. I also water the milk down until they are used to it, since it is very different from the milk they have been used to. For the first day I fed them every couple of hours. I also weighed them – they were just under 85 grams each.

Their coats shone red-gold, but they still had their eyes shut and their tails were fine and wispy. From this and their weights I could ascertain that they were probably just under a month old. It took a while before they accepted the milk without coaxing. However, after twenty-four hours one of them, a female, began to drink quite well and even moved her front feet back and forth as if stimulating milk flow from her mother's teats. The other two were more stubborn and it was a couple of days before they too took to the syringe. I tried a minute bottle with a teat attached but they didn't like this at all.

I rose every few hours through the night to feed them. As I fell carefully back into bed after each session I tried hard not to wake Iomhair. Then I felt something crawling on my arms. A few of the wretched fleas had decided to come back to bed with me. I lay there listening to the rhythmic breathing beside me, and the gentle snores emanating from the other side of the bed. There was more movement on my arms. I slapped at it gently and then felt more, until my forearms were tickling badly. It was not a pleasant sensation. Though I tried not to make a noise, it was hopeless as I was frantically dealing with minuscule unwanted bedfellows. Then a voice beside me said, 'We are not alone are we?' and we leapt out of bed, put the lights on, and threw back the duvet to search for our companions. There were only a couple, but that was enough. Though neither of us was actually bitten, it is not the nicest experience to feel these insidious little pests using you as a multi-gym. Though my resolve not to use noxious anti-flea products on baby squirrels and hedgehogs may be good for them, there is an inevitable payback. The oddest thing is that these unwanted beasts never do bite us. We don't find them on the dogs either. They quickly vanish. God only knows where. Perhaps it's better not to ask. These are squirrel fleas and they are not interested in any other hosts.

This particular trio of baby squirrels consisted of two females and a male. When orphan mammals are compromised and severely stressed it can lead to another difficult problem. Desperately searching for the comfort of their mother's teats, they latch on to the penis of one of their brothers. It's quite common in mammals, including fox, badger and otter cubs, and it can be difficult to stop it once the habit has commenced. As soon as I noticed that the females were doing this I separated the male off and put him in another basket. He began to fade and whimper in pathetic distress calls. He did not want to feed any more. I sprayed the tender area with Bitter Spray – a product to discourage puppies from destructive chewing. He was so unenthusiastic about his milk and was clearly pining for his siblings, I had to put him back in with them, and despite the spray and the fact that even my own food took on the beastly bitter taste, the other squirrels were undeterred. The entire area swelled up badly and he seemed unable to pass urine. I took him to the vet but it was too late, the damage had been done and the prescribed antibiotics and steroid cream proved useless. He died the next day. It was a stark reminder of the hazards and upsets associated with trying to rear fragile orphan mammals. Even though I am well aware of this, you can never stop yourself from thinking that you could have done better.

Positively, the females were thriving and their eyes were fully open. We moved them into a bigger box filled with branches and a small log to play on. Their acrobatics were impressive and there were frivolous periods of activity before they burned themselves out and sought the confines of their hat nest, where they instantly fell into a deep sleep. We kept them in the kitchen because we didn't want to miss any of the action and found it absorbing to watch their rapid development. Each day they became more and more agile. They were both too tame and climbed all over me. Care had to be taken not to let them cause any damage in a room full of hazards, including two dressers filled with china. It was hard to catch them when they were intent on playing, and they loved coming out into the room, where they ran up the curtains and along the window poles, before leaping onto a bookshelf. Sharp teeth can cause ruination in seconds. It was time to get them out to the aviary, but the weather was still bitterly cold.

The bad winter had taken a heavy toll on the aviary and two feet of snow had pushed areas of the roof in. I was standing on a ladder hammering staples into weldmesh when I heard spine-chilling screams from up in the high field. A roe doe had tried to jump the fence and caught her hind leg in between the top wire and the rylock. It's quite a common occurrence with deer and usually leads to a protracted and agonising struggle prior to death through exhaustion and stress. Iomhair and I puffed our way up the steep field carrying a blanket to put over her head, and wire cutters. How long had she been there? Would we have to despatch her if the leg was broken or badly damaged? She thrashed and struggled on the grass, suspended by the leg; her terror had worn it into a bare, muddy circle. We put the blanket over her head and I held her while Iomhair made a quick decision. It looked as though she had not been there long and the leg, though skinned and bleeding, was not broken. Roe deer shriek in fear and the noise of her distress was harrowing. Swiftly he managed to move the wire to free her. We sprayed the open wound with antibiotic spray as she erupted forth and disappeared into the wood. Astonishingly, she wasn't even lame – we'd just have to hope that the wound would quickly heal. At least in March there was no worry about flies and the threat of maggots in the open wound. We wondered if we would have heard her pathetic distress calls or seen her grim struggle had I not been up on the aviary roof.

Knowing as I do that squirrels are nature's great escape artists, the aviary still needed repairs to ensure there was no way out. We also had to re-cement all around the bottom to make sure nothing could dig out, or in. Some pine branches in the woods above had been brought down by the winter storms in the wild gales – these could be used to deck out the aviary with greenery to make a suitable squirrels' adventure playground. Finally, Iomhair made a porch so we could shut ourselves in before opening the inner door. Without this the opportunist squirrels could race out and over our heads in milliseconds.

Though we now had lots of squirrels coming to the garden for food, we were still plagued by two large cats. They had taken a heavy toll on the bird life round about and were also adept at killing young squirrels. Few people realise that the domestic cat is one of the red squirrel's biggest threats. With more and more cats, the problem grows. The rise in numbers of domestic cats and their hybridisation in remote areas with our critically endangered Scottish wildcat is yet another problem. Work continues to encourage all cat owners to neuter their pets. From a predation point of view, young red squirrels are the most vulnerable, particularly when they are first out of their natal dreys and are far too curious and trusting. Many meet their end this way; we had recently found squirrel remains in the garden and feared that if we released these babies in our garden, they might be feline fair game.

We put their box inside the aviary's house so they could still use it for sleeping in. I was also thinking ahead, because they would need to be transported elsewhere for eventual release. Hopefully at the allotted time we could shut them up while they slept and therefore minimise handling and the stress of chasing them around the aviary to catch them for release. Even after being hand-reared, capture can cause stress and shock.

Mark Stephen from BBC Radio Scotland's *Out of Doors*

programme came to do an interview. Cheery and full of fun, he is always game for anything new. He was visibly delighted, and almost childlike, to have not one but two red squirrels climbing up his legs and investigating the microphone. Though it proved hilarious, afterwards I worried that the two squirrels were far too trusting of a total stranger. It might lead to serious danger once they were released.

We backed right off. Other than putting large amounts of food and fresh water out every day, we left them alone. One went wild very quickly and wouldn't let us anywhere near, but the other, who had always been friendlier, took longer and still approached me. Eventually she too became wary – this was exactly what was required.

Due to the issue of the local cats, we had been thinking about suitable release sites. Cluny Gardens, high above the Tay Valley near Strathtay, was the ideal place and its owners, Wendy and John Mattingley, were thrilled at the prospect of having the two babies. A magical woodland garden open to the public, it is also very close to where the squirrels originated, and it has a healthy population of red squirrels that the Mattingleys feed every day. They are passionate about their squirrels and did not have a problem with cats, and importantly the garden is also well away from the main road.

We rose early on the last day of May. Release is a high point but never ceases to send my nervous system into overdrive – so many hazards, and as I have said before, any animals and birds that have been hand-reared are immediately starting off with a disadvantage. I had closed the door of the box the night before while the two little squirrels were asleep. The glory of a misted dawn was punctuated with a symphony of bird song, a blackbird and several song thrushes and a wren leading the medley. Beaded silver dewdrops stippled every leaf and flower, and the smoky haze of bluebells in drifts painted softly greening woodland. Cluny is glorious at any time, a haven of native plants, Himalayan and other woodland species, but it is in spring and early summer that the garden is at its finest. The Mattingleys use no pesticides and weed everything by hand. They invest heavily in the wildlife all around them and the garden is testimony to their hard graft. We were greeted by several blackbirds with beaks brimming with worms, and could hear their babies chirruping in the flowerbeds awaiting breakfast. An intoxicating scent of blossom filled our lungs as a thrush sang from an arbour above woodland paths fringed with dozens of candelabra primula. Below on the valley floor, rills of enticing mist began to burn off in the warmth of the sun.

We had chosen a release site adjacent to a giant sequoia more than 130 years old. Next to it the Mattingleys have various squirrel feeders that they keep topped up with peanuts, hazelnuts and pine nuts. Already we could see tinges of red where high in the canopy the local squirrels were stirring.

With trepidation we opened the box. In a spirited dash, the squirrels were out and immediately leapt for the great trunk before them. Having never climbed anything of either its height or girth, they made their way swiftly up into the sky above. The first proper tree they had ever climbed was this giant sequoia, a record-breaker – Britain's widest conifer with a girth of 11 metres. While we stood and watched, every now and then they would look back down on us, and then race one another around the branches.

We sat on a bench below and watched, awestruck. There are moments when it is inappropriate to speak. John Mattingley appeared a few minutes later and together we watched the squirrels. Later he put a box up in a nearby tree for them to sleep in, using the soft bedding that they were used to.

We returned several times and had regular positive updates. For a squirrel, Cluny is as close to heaven as it gets. We could do no more than let nature take over.

15

Squirrel Summer

26 June 2018

It's a glorious June morning, the snaking Tay Valley shimmers far
below, blackbird, wren and song thrush lead the dawn chorus as I
walk through the flower-stippled wood above the farm with the
three collies. Bracken is staging its annual takeover bid, racing ever
skyward, transforming hillsides to a vibrant emerald green. Small
pearl-bordered and dark green fritillaries and meadow brown
butterflies kissed awake by the breath of the sun flit back and
forth, and a cuckoo, its voice now changing to a throaty hiccup,
calls close by. It's probably the last time I will hear it this season.
Every year there are fewer and fewer of these unique brood parasites, and their onomatopoeic calls always fill me with sadness
– in my lifetime cuckoo numbers have plummeted. The familiar
sound is only heard for a short time from their arrival in April
to their departure towards the end of June. For the rest of the
year cuckoos are largely silent. As they leave after less than two
months with us, I am always left wondering, will I hear them next
year? Will they be back? Changes in farming practice as well as
colder, wetter springs with few of the vital hairy caterpillars they

need for valuable sustenance after their arduous and increasingly tortuous migrations from Africa, as well as continual habitat loss, all contribute to making life difficult for the cuckoo. Around the country the first time the call is heard has spawned a wealth of folklore and numerous bizarre beliefs. Like the also diminishing swallow, the cuckoo is a harbinger of spring; neither bird will visit a garden bird table and we are powerless to help them in this way.

The phone rings. It's a friend, Innes Smith, who is the gardener at Atholl Palace Hotel at Pitlochry. He comes straight to the point. 'Good morning, Polly, can you take three squirrel kits?' The intense love I have for red squirrels, and the pressing feeling of a need to try to help save any that are in trouble always brings an adrenaline rush, a swiftly beating heart and a shiver of excitement. As Innes continues and tells me more of the story, that moment of excitement dwindles. I fear I won't be able to help. 'I found the drey lying on the ground under a big tree in the garden a couple of hours ago. I had a very quick peep inside and saw kits. They're about an inch long, and are naked and blind. I left them where they were and watched from a safe distance in case a dog, cat or bird of prey took them. I really hoped that the mother would return, but she hasn't. In fact I haven't seen any sign of her at all so I think we need to pick them up now. What do you think?' Innes, who is a true countryman, a retired hill farmer and a man who has an extensive knowledge of wildlife, had done exactly the right thing by leaving the drey to give the mother the chance to return. But we both knew that if he didn't pick it up, the babies would perish within hours.

'An inch long? That's minute, they must be very young indeed. I am sure I won't manage to do anything with them,' I stammer, feeling helpless – ill-equipped to deal with such neonates.

'Will you take them anyway? Are you around? Can I bring them over to you now? Please will you give it a try?' He sounds desperate. We both reckon that something must have happened to the mother, as they do usually come back to carry kits to a new drey when such accidents happen. Innes tells me that he has seen a buzzard around in that part of the garden, and we both wonder if perhaps it has caught the mother and knocked the drey out of the tree. It's also not unusual for dreys to collapse, or the branches that hold them to snap and for the entire structure to hit the deck during storms or when the trees are in heavy summer leaf. Thanks to the excellent padding and insulation inside, young are often unharmed.

After our conversation I rush around making preparations for the arrivals. Frankly, I feel it's a waste of time. When Innes arrives, I am even more dubious of my chances of success. The drey, constructed of small branches and dried leaves and grasses lined with sheep's wool, is in a large cardboard box. Together we nervously and gently open it and see the minute, naked, blind, but surprisingly active kits. There are three and they are about the size and shape of a man's thumb. Their bald bodies are a pinky shade of grey, their firmly closed eyes bulge through thin membranes of skin like those of a newly hatched songster chick. Their little bodies feel clammy to the touch. They are all squirming and even at this early stage of their development seem full of vigour. I have never, ever reared squirrel kits that are this young – we estimate that they are probably no more than four days old. Surely this is going to be hopeless? What if they are actually even younger? Surely when they are this young and this vulnerable, I will have no hope. What if they have not had colostrum? Should we just let them slip away? We both note that the fleas in their drey are particularly active too.

A feeling of numb resignation engulfs me. I mix minute amounts of weak milk solution with glucose and put it in a 1-millilitre syringe. They probably won't suckle, but I suppose I have to at least go through the motions and try. It's been a stifling hot start to the day and the mercury is rising – they need liquid as a matter of urgency. Both Innes and I are astonished when they readily take the minute drops of milk. The amount I give them is critical, as at this point if I overdo it that could also lead to disaster. Too much will be every bit as bad as too little. As the milk flows through their tiny bodies, I can see as well as feel the milky pinky-white pouch of their minute tummies. From this it's very obvious that I really do need to get the amounts exactly right. With babies this small there is no margin for error – none at all.

For the first few weeks of their lives, mammals must stimulate their offspring to pass waste material. They do this through licking. Not only does it stimulate them but it also ensures the nursery area is kept clean. For the same reason many birds remove faecal sacs from their nests too. So after each feed, we must also do this. I use a warm, wet cotton-wool pad or a cotton bud to encourage them to urinate and defecate. It's a vital process, and it is extraordinary to see how well it works. I do this with newborn hedgehoglets too – indeed, any mammals. Now the squirrels are here and they have taken that first tiny feed, it does little to reassure me that they have a chance. To me at this point, it seems futile. The next forty-eight hours are critical and I will need to establish them on very regular feeds. An hour after their arrival, I give them another tiny amount – this too is eagerly taken, and now their minuscule front paws work against my fingers as I hold their vulnerable little bodies. I can't believe how keen they are to feed. It brings a lump to my throat. There are two females and a male, though one of the females is smaller than the other two, and seems perhaps a little less active.

As I have mentioned earlier, it is paramount that all newborn mammals receive their mother's colostrum. Without it the sub-stitute mother – in this case me – is going to stand very little chance of rearing them successfully. Without the vital antibodies provided by that first milk, many stress-related infections will take over and the babies will be highly likely to die as a result. Now I begin to worry about whether they might actually be so young that they have not had this. I go back to the basket to check their umbilicus again as this will give me a slightly better idea of their age. I think I know the answer already, but I am riven with the worry that perhaps I made a mistake with that first cursory glance. Did I miss something? In newborn animals the area is usually wet and bloody and often part of the cord remains before it shrivels and disappears. The three bird-like babies are closely packed together and as I gently examine them, I see that the umbilicus is indeed dry and there is no sign of even a speck of blood or the cord. It's still going to be a massive challenge but as the saying goes, where there is life there is hope, and even at this stage their will to survive is tangible.

Later in the day, after they have had a few more feeds, I weigh them. Their weights indicate that they are probably around four days old: 17, 19 and 21 grams. It's a major relief as it means that they have indeed had the magic potion – colostrum. If they live, then my summer is going to be dominated by three little squirrels. If this is to work, they will consume my every waking hour. And to begin with, most of my sleeping ones too.

The day's heat eases little at night, the bedroom sultry, sleep elusive. My head is awash with images of tiny squirrels and I rise every few hours to feed them more drops of milk, and to refuel their hot water bottle. Their eagerness to suckle and the little squeaking sounds they are making is going to make this all so much harder to bear if one, or all, succumb. I am desperate to succeed. Already I have fallen into the trap of this unexpected commitment. The kitchen windows are wide open and tawny owls call from the garden. A moth and two leatherjackets are doing a reel around the soft yellow glow of the lamp as balmy air carries the perfume of roses into the kitchen.

I am always painfully aware that just when I feel things are going really well, disaster may strike and a situation can deteriorate horribly. It's a down-side to working with wild animals and birds. I think back to my farming days, when we would walk around our ewes and lambs and comment on a particular ewe with a really good, big lamb that seemed to be thriving. Then a couple of days later we might go to check the sheep and find the lamb dead, feet in the air – for no apparent reason. People joke that sheep love to die like this. It's always heartbreakingly disappointing. With injured or orphaned wildlife it seems to occur more often, but the reasons are probably different. For tiny squirrel kits, the loss of their mother is often the unseen peril that lurks, waiting to catch the rehabilitator unawares.

The weather is extraordinary, with temperatures higher than we have had for years. Skies over Highland Perthshire are an eternal cornflower blue and for once the atmosphere lacks the usual moody humidity. It's glorious, though our sheep and Ruby are finding it hard going and lie panting in the shadows of the trees overhanging the field margins. Armies of clegs with their nipping scythe jaws catch me every time I venture out, and bluebottles erupt in a fizz from a piece of festering rabbit leg dropped by the owls last night.

After a sleepless night, I am running on adrenaline and tea. The babies have all survived and though the smallest seems frailer, they are surprisingly enthusiastic about their feeds. I have now added a tiny teat to the end of the syringe and they have quickly adapted and are suckling enthusiastically. I know they would suckle a bottle but I am aware that they will drink too fast in their keenness to feed and perhaps take too much, or worse still inhale it. By using a syringe instead I can control the speed of flow as they drink.

When the babies first arrived, I had been unable to get hold of the preferred, and best, substitute milk powder. Over the years brands have evolved and improved and it's always useful to be able to speak to fellow rehabilitators to see what has worked best for them, what they are using and recommending. This time, I have had to use a kitten milk substitute instead, and though it has worked well in the past, I know that they will fare better with a higher fat content, particularly being so young. Changing milk is generally to be avoided. However, my friends and mentors at the Scottish SPCA's National Wildlife Rescue Centre, Sheelagh McAllister and Colin Seddon, and world-renowned wildlife and exotics vet Romain Pizzi, advise me that I need to put my tiny squirrels on to the better milk powder as soon as I can get hold of it.

Day three, and all the squirrels are doing extremely well. I have to feed the little one twice during a feeding session – a tiny amount when I start the round of the three of them, and then again at the end. She is not such a good feeder and is slower and needs to be coaxed. Still lacking any hair, their little bodies are always clammy to the touch and I have to take care to keep them really warm as I feed them. They chill at a frightening rate. Every spilled droplet or splash of milk must be carefully wiped away too in order to prevent it hardening around the thin, tender skin where it might cause sores. Toileting takes place at every feed and then I wipe each baby with a warm cotton-wool pad to ensure it is perfectly dry, since urine can burn the skin. When they are all full and I can see that little visible white pouch of milk in their stomachs, I stroke them with a soft squirrel-hair paintbrush. They curl up chortling and squeaking, contented and replete. My thought is that if they are relaxed then hopefully they won't be so vulnerable to stress-related issues. Their mother would be licking them thoroughly at this point.

I have been weighing them regularly and they are all putting on weight. I am loth to muck them about now with a different milk powder but I know I have to do it. I tentatively make the change over the next twenty-four hours. They are totally unenthusiastic about the new milk – it's far thicker and more gloopy, and even though I am mixing it gradually with the other milk, it's obvious they don't think much of it. Now they suffer a worrying setback and all three lose a little weight. 'Persevere, it will be fine,' both Romain and Sheelagh assure me. I feel lucky to have them at the end of the phone. Sheelagh has reared numerous mammals, including stoats, weasels, pine martens, foxes and badgers, though she says she has never had squirrels this young. She has a natural affinity with them, a gift when it comes to animal husbandry.

16

Three Squirrels in a Bookshop

The next twenty-four hours are sore on the nerves. I have a new book – *A Richness of Martens* – coming out in a few weeks' time, and plans and preparations now have to be slotted in around squirrels. It's also increasingly hard to concentrate on my usual work for the squirrels are taking priority. The small kit has little energy but the other two are doing very well. The milk saga eases and after riding another emotional roller-coaster, calm is restored. Once again they are all feeding happily.

Rising a couple of times in the night is no hardship. The kitchen is otherworldly, its windows flung wide in the hot breath of summer. The rhythmic heartbeat tick of the clock, the churring of owls and the contented sounds of three little squirrels. Bats pass the window, and I think I can hear a badger snuffling on the bank. Far away a dog barks and then is silent; there is only stillness on the warm sigh of a ruffling breeze.

And then at the last feed before bedtime, the littlest squirrel has too much and when I return in the middle of the night I find her cold and listless. She won't feed and I am sure she is going to die. I spend time massaging her tiny swollen tummy while trying to get her to pass waste material, to no avail. I return to bed with a heavy heart, blaming myself for being careless. Perhaps I did indeed over-feed her. At dawn she is still cold and lacking energy but after a while I manage to get her to take some glucose water, and also to pass some waste. By the next feed she is back on track but I take extra care and instead of feeding her three-hourly, return her to tiny amounts every hour until her little stomach has settled again.

I realise at this point that if even one of them dies I am going to be devastated. I am trying to keep my distance but it's impossible when hand-rearing such diminutive animals, impossible when you invest so much energy, not to invest emotion too. Impossible. I fear I am far too emotionally embroiled. I must take a step back. I have to keep things in perspective.

The situation is not helped by the fact that Innes posted images of the squirrels in their drey on social media after he had brought them to me. The squirrels have already gathered a following and so many people are invested in their cause. I am getting texts and messages constantly asking me how they are faring. I am doing regular updates on Facebook – something I don't usually do. Despite the fact that people are willing the little squirrels on and all mean so well, it does add more pressure. The squirrels' lives have become public.

A few days after the milk setback, there is a suggestion of soft ginger fuzz sprouting like orange grass seed all over the babies. Their ratty tails are fluffing out too. After their feeds, they love their gentle brushing sessions even more and nestle into the palm of my hand before going back to sleep. By the time they are five weeks old their beady little dark eyes are beginning to open and I see the biggest female eye to eye for the first time. Her eyes are the colour of brambles. Suddenly the babies are more alert and ready to play for short bursts when, like engines at full throttle, they race around, and then their system floods, and sleep is all they want. I know the feeling. I once heard this referred to as shrew metabolism for it's what shrews do too – frenetic bursts of hunting and feeding and then they literally seem to burn out.

The kits' supremely adapted hind feet continue to appear far, far too big for them. It's something I always marvel over with all the squirrels I receive, even adults. That special adaptation of double-jointed ankles and those incredibly sharp claws contribute to making the red squirrel nature's Olympic athlete. Their new coats gleam as if they have been varnished. Despite the setbacks at the start, the new milk is noticeably superior. Sheelagh did say to me that she had noticed that her squirrels seemed to have shinier coats than when she used other brands.

At this stage, weight-wise the male is always slightly ahead but relatively all three are growing at a similar rate. Freddy is making a fascinating graph of their growth rates. He has called it Squirrel Tracker. This gives us a clear indication of how each is progressing. It is interesting to note when there are dips in their weights, indicating I need to up their feeds. Soon I will be introducing them to a little solid food. Now I am dreaming of eventually releasing them here in the garden since the two cats that were causing us concern are now too elderly to bother with hunting, and two neighbours have moved and taken their cats with them. It's been my goal since the start, but together the little squirrels and I still have many hurdles to jump. Is it too much to allow myself to aim so high?

Due to the presence of the squirrels many of my plans have to be postponed, or shelved altogether, yet it still remains an extremely busy spell. The weekend of my new book launch looms ever closer. Then we will have a house full and are having a party in our big shed. The babies have not had visitors as when the aim is to return animals to the wild, the fewer people they see and become accustomed to, the better. Tameness is detrimental in the wild. It's also important to recognise that they are not exhibits, and anything that might risk stressing them has to be avoided.

The heat continues; the garden wilts. Usually in high summer showers of monsoon proportion bring flash floods and increasingly dominate the Highland summer and flatten my tall herbaceous plants. Now instead the blooms lie horizontal, succumbing to intense thirst. Watering is an endless round that seems to offer little relief. Scotland is a paradox. Over the big weekend I shut the squirrels away in their basket in my office upstairs where it is quiet. I creep away at regular intervals to feed them, and find the routine helps to keep me focused. There is stillness and peace taking time out to care for them, listening to their little noises, as they take their feeds with gusto. Later we note from Freddy's Squirrel Tracker graph that over that particular weekend, though I thought I had given the squirrels the same amount of care and attention as usual, their weights remained static. Once the weekend is over, they begin to blossom again. It proves that all my mad rushing about has clearly rubbed off on them. Animals, even at this tender age and stage, are incredibly perceptive.

*

There have been several occasions where animal-minded people have contacted me and said that they would love to get into rehabilitation work. It is something I am keen to encourage – there are probably far too few of us around the country. Humans are indeed the cause of most of the problems associated with the work, and though success rates are low, every little we can do to help rectify the problems we have caused is surely a good thing. However, as I explained earlier, most of those who have started out enthusiastically fall by the wayside – and usually after a woefully short time.

Living where I do in the very heart of Scotland we are equidistant from the sea on both sides, but it's not unusual, particularly in summer following storms when the coast has been savagely gale-lashed, to receive calls about 'little penguins'. It has become a joke in this family and with the local vets too, who regularly receive similar calls. You'd be forgiven for thinking guillemots look penguin-like – indeed the French call the closely related and similar razorbill, *petit pingouin*. Occasionally young guillemots are disorientated, having been blown off course as they try to venture out to sea from their natal cliffs. Some of them land in gardens, or are seen bobbing around on Loch Tay, far, far off their destination. Usually they are starving hungry; they will quickly recover on a diet of oily fish and can soon be taken back to the coast. On one occasion a woman rang about one such casualty and told me she had always wanted to work with wildlife and would like to look after the starving guillemot herself before returning it to the wild. She asked me what she needed to do. Typically, her new-found interest in rehabilitation didn't last long. A day later she rang again and told me that she had visitors and really didn't have any spare time for looking after the bird at the moment. 'It also smells terrible, so I am wondering would it be OK if you took it? It's really not convenient for me at the moment. Can I ring you again in future and see if there is anything else you would like help with as I would very much enjoy to do this work again.' Like help with? Really? It's hard not to be sharp in response. I sigh to myself and make a mental note that she won't make a rehabilitator. The point here is that casualties never, ever appear when it's convenient. Over the years I have had to take numerous birds and beasts with me on work trips. The timing of the squirrels' arrival was certainly not ideal. But this is almost always the case, and if you are serious about helping wildlife it's something you have to work around. And there are not that many people, unless they are paid to do the work, who can afford to stop the day job.

*

Following the book launch at the Aberfeldy Watermill Bookshop, I have a trip later in the week to Fort William, to the Highland Bookshop for an event on my book. After this I am going to Ardnamurchan for two nights to hold an event there too, as much of my book is based in the area. I will be taking the squirrels. The Highland Bookshop event is in the evening and, following military-style planning, I set off with my precious cargo and all their gear. They will be oblivious to the entire foray. I gather together all their accoutrements to ensure I don't forget anything: clean bedding, milk powder, soft brush for comforting sessions after feeds, syringes, spare teats, sterilising fluid, cotton-wool pads, hot water bottle, thermos flasks of hot water – for topping up said hot water bottle if I should break down on the way to Fort William. Then there's all my own gear to think about too. The timing is working out fine but if the event were even a week later, the squirrels would be far too active to make it possible, as the risk of their escape would be high. Their enhanced awareness might risk stress to them too.

Iomhair waves me off, 'It's worse than having a new baby,' he laughs, seeing all their equipment piled up in a large bag. The drive west is one of my favourites. There is the strong feeling of returning home. The afternoon is golden: lochs mirror hills and glens, rowanberries starting to turn orange by the roadside, the blush of pink and red foxgloves, heather and rosebay willowherb. Thin, wispy cloud embraces mountain summits, and a journey that is often atmospheric and moody, lost in mist and yes, interminable rain, instead basks in warm light. Glen Coe is benign, its grey summit ridges scoured and scored by centuries of weathering and boot traffic. Ossian's Cave awakens a memory of the very first time I glanced up and saw its great gaping mouth. I was sitting in the back of my parents' car as we drove west to see the Kilchoan Hotel, where they subsequently moved. The dark opening in the rock face of the Three Sisters peak on Aonach Dubh filled me with intrigue. It was straight out of a fairy tale, a place of dragons and fire breath, of Andersen and Grimm, of evil deeds and magic, as well as of refuge and shelter. To a child, a cave is never simply a cave. Glen Coe, the Lost Valley, has held me in awe ever since, with its reedy roadside lochan where the graceful snowy forms of whooper swans are often seen in winter, pristine spectres shining from a palette of greys, and the distinctive shape of the Pap of Glen Coe looming over Loch Leven at the western end of the Aonach Eagach ridge.

As I walk down Fort William High Street late that July afternoon, would anyone suspect that what I am carrying – a lidded picnic basket – actually contains squirrels? This is surely the first and only time red squirrels will have had their milk in secret in a bookshop office, to the delight of the shop's effervescent manager, Sarah-Lou Bamblett and her colleague Kelsey Ward, who sneak in for a quick look as the trio have their supper.

After the event, I feed the babies again before catching the ferry across Loch Linnhe to drive the long, twisty road down the Ardnamurchan peninsula. Now I really feel as if I am going home. After such an exciting day and with my new book hot off the press, I am over-emotional – so many memories of my family, my unsurpassed childhood here, and this peninsula that has without doubt been responsible for my way of life as a wildlife writer and naturalist. It's here that I have had some of my finest wildlife encounters, and continue to do so. Every bay brings another vibrant memory: otters, pine martens, seals and eagles. And herons standing silently in wait for fish. Volatile burns bouncing and bubbling their way off steep verdant hillsides, and oak woods with their cushions of vivid emerald mosses, ferns and lichen-crusted rocks sprawling low, shaped and honed by the rigours of the gales that sweep in off the tempestuous Atlantic. I stop in a passing place on the side of Loch Sunart to inhale the fading light over hammer-marked grey water tinged with an apricot after-glow. A buzzard mews in the distance and I see silhouettes of hinds and calves on a west-facing knoll. The squirrels sleep on in their basket. It's nearly time for another feed.

17

Squirrels Go West

As I round the familiar bends on the approach to Glenmore, a smirr of rain stipples the windscreen as two hinds with their calves cross the road in front of me. I turn into the driveway and realise how tired I feel. Midges have mustered their armies and manage to find their way into my jacket as I go back and forth ferrying the babies and all their gear into the house.

Les and Chris Humphreys are two of the finest hosts I know. The extraordinary story of these retired dairy farmers, and the delights of their astonishing wild garden on the shores of Loch Sunart on the fringes of Ardnamurchan's frayed Atlantic oak woods, forms the main thread of *A Richness of Martens*, the book I am here to promote. They have turned their garden over to wildlife entirely. Using camera traps, as well as CCTV, every corner of their property is bugged, enabling them to capture unique footage. Their detailed records of all the pine martens they love and know so much about, and that have regularly visited their garden over the past fifteen years, are a valuable record and reveal many unknown facets of the marten's character, and its natural history.

From otters to foxes, badgers, hedgehogs and herons, to the annual frenzy of frog fornication in their garden pond, every movement is captured in this wildlife paradise, and Les and Chris are constantly discovering something new. Squirrels, however, as I explained earlier, are not present in this part of Ardnamurchan, though they are making a sporadic return to areas further east on the peninsula. Visits to Les and Chris are always joyous and I find it hard to tear myself away from this haven of wild intrigue. It's a place like no other, and as I sink into one of their capaciously comfortable armchairs with a mug of tea and some of Chris's unsurpassed baking, I find myself watching otters, seals and porpoises from the large window over Glenmore Bay on Loch Sunart, and pine martens either on their CCTV, or better still in the sitting room itself, when some venture in to gently take an egg from Les's hand. On these occasions, it's impossible to know where to focus attention.

Les and Chris are passionate about all animals and they greet me effusively. They are dying to see the squirrels and to ensure that they are settled before we do anything else. Animals always come first.

It's late and the first priority is to feed the babies. Chris makes tea and we sit around their kitchen table listening to squeaking and contented sucking sounds as the ginger trio relish another meal. Les and Chris are delighted. Chris tells me about marmosets that they once hand-reared. They sound remarkably similar and required the same routine and dedication. Chris could successfully hand-rear anything. I pass her one of the little squirrels when it has been fed and she uses the soft brush to soothe it. After initially investigating this different hand, it curls into a tiny ball and squeaks in appreciation like a purring cat. We repeat the process until they are all fed, refill the hot water bottle and return them all safely into their hat.

The kitchen is still, rain gathering momentum. Most of the regular visiting pine martens came earlier, but there are rumblings outside in the garden. A badger is proving tenacious and defying all Les's attempts to stop it devouring the food left out for martens and hedgehogs. We sit chatting about the latest visitors to the garden, pine martens I have seen here, and some that have been coming for many years, new ones that have been seen recently, and others that have simply passed through. Marten gossip! It beats listening to what the neighbours have been doing.

I descend to the kitchen in the small hours to feed the babies. The forecast is for horizontal rain all day and the three of us are to spend it in a marquee at the Kilchoan Show to promote the book. I am swiftly going off the idea.

None of us are very enthusiastic as we leave early to set up our stand, but we have made this plan and have to do our best. There will be lots of visitors around and as the book is also about Ardnamurchan and its wonderfully diverse wildlife, we hope it will prove popular. I have boxes and boxes of books in my car, and Chris has a car-load of sumptuous refreshments to keep us fuelled during the show. I am sure it will be one of those non-descript days where we comfort-eat without cease. The squirrels will stay here and I will return to feed them. I don't want to risk taking them and trying to feed them in the marquee. It's safer if they stay put, but leaving them makes me uneasy. What if I cannot get back in time? I still feel it's safer and quieter here.

Kilchoan is lost in low cloud and rain, many of the show events cancelled. We set up shop in the dripping corner of a draughty marquee. 'We probably won't sell any, it's so hidden away,' says Chris gloomily, voicing my own fears. It turns out to be a miserable day – few people come near, and seem to bypass us in favour of the beer tent. And though a handful of loyal locals purchase the book, the day is dominated by endless negative comments. 'Bloody pine martens, I hate them, they kill hens.' Or, 'There's far too many pine martens, they attacked my cat.' To this last I ask, 'And did you see them attack your cat?' 'No, but I know it was those vicious little bastards, they murder everything.' This cultural dislike of predators gets me down. It gets Les and Chris down too. None of us are under any illusion that pine martens are angels, and of course they kill hens and filch eggs, but it's important to reiterate once again that we need predators as we need prey. Nature is out of balance. However, these are dyed-in-the-wool characters who won't change their opinions now. It's a waste of time explaining. Better to save the breath and smile, 'Yes, you are quite right' through gritted teeth. We stand there feeling cold and tired, and wishing we had never embarked on this. We graze all day, consuming lots of Chris's rolls and cake. We drink tea. We shiver with our feet cold and water seeping down our necks as we make polite conversation with the few who stop by. Then the subject of sea eagles and sheep conflict comes up. I knew it would. And I am accosted, trapped under a waterspout cascading off the corner of the marquee. 'There's far too many bloody sea eagles, Polly. They take lambs. We hate them here. And they also threaten our golden eagles.' That last is complete nonsense but is something I hear quite often from those who hate sea eagles – ironically, they hated golden eagles before. How many is too many, I wonder. And ask with my wet neck, wishing I had one of brass instead. I don't even bother to defend the local wildlife, or talk about nature's balance. It's a lost cause, and as my mother would have aptly said, 'It's like farting against thunder.' But the few tourists who have come to the washed-out show are different and buy the book with enthusiasm. Thank God!

And in between I drive all the way back up the winding road to Glenmore, to the peace of the house by the shore, where I feed my babies. Book promotion can be so hard at times. By the end of the day we have sold far fewer than expected and our confidence has been severely dented. We feel punctured. Bloody pine martens – and the final comment – 'Well, they eat red squirrels too, you know.' And that brings me back to my point that less than a century ago you would have heard a similar comment, 'There are far too many bloody red squirrels, so they must be culled.' And look at how we view squirrels now. Our attitudes are changing, but getting there is like pushing rocks uphill through glutinous mud in an overladen barrow with a dodgy wheel. And that thought makes me giggle.

Glenmore is a haven where we spend a glorious evening together watching the coming and going of the 'bloody' pine martens. They are beautiful. Sleek in their summer pelts of rich chocolate, gleaming like well-polished mahogany. The badger comes too and causes me much hilarity as it bombastically fails to be bamboozled by Les's patent hand-made barricades, and bulldozes straight through them. Bloody badgers!

Next day Glenmore Bay is a cauldron of boiling activity, the water electrified by thousands of silver sprats and mackerel. Herring and black-backed gulls, as well as a few gannets, have congregated and are taking advantage, diving and splashing, re-emerging with gullets bulging. The shore is lined with herons – there are dozens, all feasting on the fish takeaway. We watch from the window as squally showers win over brief breaks of light, and the bay colours with rainbows. We joke that perhaps this is the herons' AGM as more and more drop in, their long legs outstretched, prehistoric forms against the sky. And we joke about far too many bloody herons. The truth of the matter is that there are far too many bloody humans.

As we feed the squirrels before I leave, we notice that there is an odd whorl of hair around the largest female's nose. It looks as though there is a lump developing. I will have to keep an eye on this. They are doing so well, I don't need any setbacks now.

Once home, the lump grew steadily bigger, though it didn't appear to be bothering her. I also noticed that the small female had one too. They looked like abscesses, but what had caused them? As I have mentioned, squirrels have very sharp claws, and on arrival the tiny naked babies had noticeable scratches on their faces, perhaps caused by their mother in the drey. I wondered if these had led to infection. I contacted wildlife vet Romain Pizzi, who works with the Scottish SPCA at their wildlife unit and whose skills are much in demand throughout the world. His almost instant text reply made me smile – he was in China operating on a giraffe. As ever, Romain restored my faith in humanity, for he was as equally concerned about my squirrels as he was about the giraffe. He asked me to email photographs, then diagnosed abscesses and prescribed paediatric banana-flavoured antibiotic syrup. Within days the abscesses had burst spectacularly with no ill effects. Thanks to him we had overcome another hurdle.

18

Letting Go

I have been working closely with rewilding advocacy charity Scotland: The Big Picture (SBP), whose aim is to promote a wilder Scotland, a place where both wildlife and people thrive. I agreed that providing it was not too intrusive, members of their team of media professionals could make a short film documenting the story of our three squirrels. If the squirrels with their indescribably powerful charms could help to highlight the urgent need for connected habitat, together with the vital importance of native woodland, then it will be a job well done, for collectively this is our aim. Using a mixture of their own filming and clips that we are taking, the film will tell the squirrels' story and hopefully also include their eventual release.

Filmmaker and SBP director Mat Larkin comes to film them on several occasions. He is patient and undemanding, and understands perfectly that animals don't do things to order. On his second visit, the squirrels are more interested in investigating him and his vast array of heavy filming equipment than being the stars of his show. They bounce in and around the kit bags, and leap from me to Mat. One of them perches precariously on his head while he is trying to film me feeding one of the others. It's not the easiest task to work out their movements and keep track of the three miniature whirlwinds as they speed around the room in a totally random way.

These crazy, energy-packed little mammals are uncontrollable. Having them racing around the house is firstly unsafe for them, and secondly totally impractical. A bull in a china shop portrays a distinctive image of destruction but even little squirrels can cause mayhem and not only chew everything, but also send things flying as they zip around the room. It's definitely time to move them out to the aviary.

Like others before them, they are also far too familiar with us. Though it's inevitable under the circumstances, it is another concern, and now we must address it. I knew from the minute they arrived that if they did survive, keeping them in captivity would never be an option. They need to be free, and I won't have succeeded if they cannot be returned to the wild. However, as they are the youngest I have ever reared, I cannot help wondering whether it might be harder to get them away than it has been with others. Will they keep jumping all over us? Might they leap on casual passers-by on the footpaths around the area? If so, it will lead to disaster. All the squirrels I have reared have quickly become wary as soon as they are left alone in the aviary, where I am careful to minimise contact. Surely it will be easier with three of them, because they have each other? My head is filled with doubts and trying to reassure myself seems pointless. I can argue myself into a hole and back out again with my overactive mind.

We line one of the aviaries with new perches and climbing logs, cover the floor with dry leaves and hide nuts, cones, seeds and berries. Iomhair makes them a lovely new nest box that I stuff with sheep's wool, leaves, lichen and moss. We carry out their box cage while they are asleep and carefully put it in the aviary's house. If needed, they can return to this until they are accustomed to the

nest box. We also put up a nut box feeder. Before release they need
to know how to use one so that they can return to the garden at
any time for food. In recent years the cat threat has diminished
and we have fewer visiting the garden through the day.

On the first night I find two of the squirrels fast asleep inside
the box feeder on a pile of hazelnuts and peanuts. It looks a hard
bed to lie on and offers nothing like the comfort they have become
accustomed to, but they are like kids in a sweetshop, having never
seen such a glut of nuts. Despite the lack of luxury bedding, they are
oblivious, clearly exhausted by the excitement of being out in a big
space. They are still curled up inside when I rise at dawn. They peer
sleepily up at me when I go in. It must have been a most uncomfortable night. Clearly, I needn't have worried about them learning to
use this feeder, they have mastered it already and gorged themselves
on nuts in the process. For the next few days they sit inside filling
themselves up greedily. And then the novelty wears off. And they
have learned that Iomhair's nest box is a far cosier venue for sleep. To
begin with when I go in to check them, they still use me as a trampoline and climbing frame. Despite the amusement and fun of being
climbed on by squirrels, this has to stop if they are to stand a chance
of freedom. Now I must divorce myself from them and only go in
to put out fresh food and water while they sleep. I will check them
from a safe distance to ensure that they are all continuing to thrive.

We watch the babies racing around their new abode from the
kitchen window. Growing very fast, their food consumption
is high, though like most squirrels they are incredibly wasteful,
nibbling a little bit of this and that before discarding it to try something else instead. Blue tits and coal tits as well as a robin squeeze
themselves in through the small-gauge weldmesh and feast on the
bits that the squirrels disdainfully chuck out, and there are little
brown wood mice taking advantage of the situation too. They sit
up with pieces of fruit in their tiny front paws and nibble daintily

before scurrying off into the leaves, but they are wise enough not to venture into the adjacent aviary, where there are owlets!

Over a month passes with the squirrels in the aviary. Release day is approaching. We are nearly there and I cannot believe that together we have reached what seemed such an impossible goal. I am filled with so many mixed emotions. They are so small and vulnerable, yet I am seeing far smaller squirrel kits out and about on my daily walks, and yesterday as I walked through the nearby village of Weem and crossed over to the riverside path, a tiny squirrel kit raced along the wall by the roadside – it was far younger than our three and already out and independent. Our babies have to take their chance; there will always be a back-up for them with a constant food source left all around the garden.

Thankfully, the lack of contact is working and they have become wary of us. They dart off into the nest box when anyone approaches the aviary. This is an excellent sign. This will hopefully give them the best chance of survival. Tough love for squirrels. And me!

6 September 2018

It's 5.30 a.m. on release day. Mark Hamblin of SBP is coming to film the squirrels as we set them free. I cannot sleep. I love this hour of the day and enjoy rising early – I always have. This is nature's time – the garden is awake, blackbirds pulling fat worms from the grass and bats streaming gradually back to their bed under the old bothy slate roof. Tawny owls are still around and call to one another across Kim's Wood. And wood pigeons mourn the passing of summer with soft coos amid floating thistledown. Mist hides the treetops and I stand at the kitchen window and watch both the wild squirrels, and the trio in the aviary eating their breakfast. The three babies play together, and as I watch I feel both euphoric and apprehensive. So many hazards await them. Just yesterday a very large male pine marten visited the peanut feeders close to the aviaries. Though usually the martens and squirrels take no notice of one another, the bigger beast still looks threatening. His presence made me more aware of the squirrels' vulnerability. The local cats are far more of a threat, as are dogs on the path that comes close to the farm. I am reminded of the hard fact that a large percentage of young squirrels don't make it through their first winter. But this is no reason to keep them in captivity. I have to keep myself focused.

Mark is here by 7.45 a.m. but the squirrels, having been manically active since daybreak, have had a big breakfast and retired exhausted to bed for a long lie. We don't want to force them out and will only open the aviary door once they are all awake again. After he sets up his camera we sit and chat over breakfast with an eye on the aviary. Then I have a brief glimpse of an unusual visitor. A glorious hummingbird hawk moth feeds on the buddleia by the aviary. I rush out in the hope of getting a close-up view, but it has vanished. It's only the third time in twenty years that I have seen one in the garden. Well-named, it closely resembles a tiny hummingbird as it thrums with continuous wing beats, adding a touch of the exotic to my Highland Perthshire garden.

The squirrels are awake and racing around again. And in my stomach butterflies are racing around too. The moment has come; this is what it has been all about. We have reached the end together, these little squirrels and me. I go over and open the aviary door. Then stand back and wait. Often it takes time for animals and birds to come out in such situations. Sometimes they don't even notice that they are free to go. One such tawny owlet in a batch of five in the aviary took three days to come out, and I had to shut the door again each night for fear of a cat, pine marten or fox going in and catching it unawares. I never chase things out because it is always far better for them to go in their own time, when they are ready.

It takes some time for the squirrels to realise that they can come out. Mark is patient and stands watching silently. It is the smaller female that emerges first. Slowly at first, very slowly, and then she races onto the aviary roof, finds a dish of food, and begins to casually nibble some pieces of melon. Soon after, the second female joins her. They find the antler that Iomhair has fixed to the aviary and begin to test it with their sharp teeth. Then they chase one another around it, and nibble a little more. Soon afterwards the male comes to the doorway but takes longer to come right out. I have a huge lump in my throat. I am trying not to cry, for it is indeed a great moment of euphoria. We have reached our goal. But the highlight comes a few moments later and then I admit I do cry, and I hope Mark has not captured that private moment on film. The three of them race one another up and down the tallest tree in the garden, an aspen that I planted eighteen years ago when I first came to the farm. It is as if these fabulous honed athletes, leaping and bounding as if they really were flying squirrels, have done this a thousand times before. Round and round they chase, leaping through thin branches as the leafy canopy embraces them with open arms. 'That went well,' says Mark in priceless under-statement, and he smiles. I think he knows exactly how I feel.

The babies continued to return to the garden for most of that autumn. At first coming at least three times a day, gradually their visits lessened until it became increasingly hard to know which squirrel was which, but we believed that they continued to flourish. As well as ensuring that the dedicated squirrel feeder boxes were always replenished I left fruit out on the top of the aviary for them, in particular some tiny pieces of their favourite melon. Most days I saw one or two of them sitting in the dish eating it. Increasingly they came alone and separately, and usually outwith the times the other local squirrels visited, presumably to avoid being chased by them. That year our local autumnal woods were rich with natural food and gradually as the squirrels disappeared further away to explore, the local bird population appeared to become addicted to melon.

19

Owls in the Chimney

The tawny is our most common species of owl. Also referred to as the brown wood owl, it seems to be one of few birds that have successfully adapted to living alongside man. This is no easy feat. I have always had the greatest admiration for this glorious owl, with its classic hooting call so often heard as the backdrop to film, TV and video clips – the call we associate with owls in general, even though owl species all have different vocalisations. 'What does the owl say?' we ask small children. 'Tu whit tu woo,' they giggle in response, but for the tawny this is just one of a range of sounds. When you listen a little harder, you may also hear the volley of calls between males and females, as well as those made by owlets.

Tawnies are also fascinating due to the dramatic variation in their cryptic plumage. While some may be a rich russet hue, others are a palette of soft greys and browns that provide them with perfect camouflage. And we love them too because they have large dark, forward-facing eyes that make them appear almost human. Underneath this appealing guise is a winged killer with talons as sharp as any cat's claws and hearing so acute that they won't miss even the whisper of a vole in long grasses.

However, unlike other owls, such as the closely related long-eared and short-eared, because the tawny has adapted to living in such close proximity to humans, it frequently comes to grief. I have had dozens of tawnies in my care, and have hand-reared more of them than any other birds. They are usually fairly easy to treat and if I can get them to eat, usually their response to treatment is good; it makes working with them feel rewarding. Nocturnal hunting along roadside verges for small mammals can lead to disaster for the owls. Casualties may be too badly injured and a decision must be swiftly taken on the future of an owl that has broken wings or legs. If it will never fly again, then there is an inevitable outcome.

Often the owlets that I receive, like some squirrel kits, have been the victims of woodcutters' chainsaws, while some of the adult tawnies that I have successfully returned to the wild have survived horrific ordeals. One went through almost the entire process in a pea cannery and only narrowly avoided ended up with the peas, in a can. Another was hit on the road close to Norwich one dark winter's night. The driver saw the bird in his headlights and knew that he had then hit it. Despite pulling over onto the hard shoulder to look for it, unbeknown to him it must have been embedded in the front grille of his lorry. It was then discovered by a mechanic on arrival at a depot in Edinburgh next morning. Miraculously, nothing was broken but the bird was severely bruised and in an advanced state of shock. It was astounding that it had survived the terrifying ordeal of being a figurehead for almost 400 miles on a bitter winter night. I felt euphoric when I released it a month later.

Sometimes owls are snared in garden and tennis court netting, and they often get stuck in chimneys or wood-burning stovepipes. It's not always easy to extricate them from chimneys, and then once safely out, they may require a full shampoo and set, complete with blow-dry in order to remove glutinous, tarry soot plastering their beautiful feathers. For a bird to survive in the wild its feathers must be in immaculate condition. Tawnies frequently choose chimneys as nest sites. When they're no longer being used, they make great places in which to rear young, but when they are still in use, obvious dangers lurk. Unlike jackdaws, which also favour such places, owls don't make proper nests and therefore don't fill up these tight spaces with sticks, but this makes it harder to know they are there.

I don't have a favourite bird because I love so many, but there is no doubt that the tawny holds a very special place in my affections. It, like the red squirrel, has been a part of my psyche since childhood. In *The Tale of Squirrel Nutkin*, Beatrix Potter's portrayal of Old Brown on Owl Island, though anthropomorphic, reveals much about owl behaviour, in particular the manner in which owls roost through the day, brilliantly camouflaged against the bark of a tree. Her illustrations depict the manner in which owls narrow their large dark eyes, in the hope that this will make them even less conspicuous.

A large russet-coloured tawny owl was one of the first birds that intrigued me as a child. I was outside playing before school when I heard a soft hooting, trilling sound coming from the old crumbling stable block next to our house. That first glimpse is still clear in my mind. Perched up against the flaking sandstone walls of an old window, the owl pulled itself up tall, closed its eyes to mere slits, and lent into the window space. When Mum shouted for me because it was time to leave for school, I made a fuss about not feeling well, hoping that I could stay at home to watch the owl. Mum laughed about this and often trotted out the story. Apparently I was good at making excuses if I didn't want to do something; when it related to school, that was often. She too was a child of nature and had also disliked school intensely. As the squirrels would do a few years later when I was sent away to boarding school, that first owl consumed me and every day when

I returned from school, I raced out to see if it was still there. Then I discovered a heap of woolly grey owl pellets and I eagerly took them apart to examine their contents using a magnifying glass to reveal the owl's diet. The perfectly intact minute whitened vole jawbones complete with teeth, or tiny bird skulls, would be carefully stored in matchboxes and taken to school for the nature table.

It seems appropriate that in *The Tale of Squirrel Nutkin*, two animals that have since been a major part of my life should be portrayed together in Beatrix Potter's beautiful illustrations. Though Old Brown attacked the irritatingly teasing Squirrel Nutkin and bit off his tail, I have kept tawny owls and squirrels in adjacent aviaries and neither seems bothered by the other. It would probably be unusual for a tawny to take a squirrel in the wild, not least because the two are not usually out and about at the same time. However, I would never rule it out. Tawnies are adept hunters and a diminutive, weak baby squirrel would provide an easy target, but I think it's rare and I have never witnessed it. On the other hand, buzzards do occasionally prey on squirrels, but they are not the most athletic raptors. Goshawks – stealthy, agile and swift – on the other hand, regularly prey on squirrels. As our wildlife is compromised by loss of habitat and food, who knows what new behaviours we may witness?

The first tawny owlet that I hand-reared became too tame; he became imprinted on me, and was therefore unfit for release. Unoriginally christened Sage, he arrived in a Pampers nappy box. He was smaller than my fist and coated in grey down. His typical pink eyelids with their thick wispy fluff made him look like a wizened old man. All tawny owlets have these pink extremities around their eyes. And all owlets also have pink, bald elbows from the way they hunker back in their uncomfortable nest scrapes.

Sage was easy to rear and ate readily from the outset. I learned a great deal from him and realised that raising several owlets together would prevent imprinting. Due to his tameness, Sage remained with me for his lifetime and though the natural lifespan of a wild owl is seldom more than five years, and often far less, he lived for eighteen years. For a while he shared the aviary with an owl that was left to me as an unusual legacy in the will of an elderly friend. Even more unimaginatively named Owly, she had a permanently damaged wing and couldn't fly. Sage and Owly immediately paired up, and for several springs reared a single chick, until Owly died at the age of twelve. Their single owlets were always good doers and emerged from the nest box looking fat and prosperous. This was no surprise given the amount of food their devoted parents fed them via me. And each year I released these tubby owlets and continued to put food out every night for them, as I do for all the owlets that are released from the farm.

Usually, owlets come back nightly to take advantage of the free takeaway; a few, like their wild-reared counterparts, may fall by the wayside, but in general they do well. Mortality in young owls in the wild, as for squirrels, is surprisingly high, and more than half of them won't make it through their first winter.

I have been asked whether I think I should be feeding the owls that frequent the district. We feed all the small birds, and put food out for squirrels every day. Sadly, in an environment where an abundant supply of natural food has dwindled drastically in the last fifty years, putting supplementary food out is surely of benefit. Perhaps without it we would have even less wildlife. In our garden we feed year-round, though I know some people advocate only doing so in the winter.

A couple of years after I had arrived at the farm, I was digging the vegetable patch when I heard soft hooting coming from the old tumbledown bothy adjacent to the house. Though I investigated the precarious upstairs of the building and searched in the cobwebbed rafters, no sign of owls could be found. At this stage the building still had an upper floor. The farm's previous owner had used the upstairs area as a henhouse and there were piles of desiccated rat droppings covering everything. The old stone walls and crevices were festooned in dusty cobwebs spun by huge spiders. As beams of sun trickled in through gaps in the slates, they illuminated ancient hen skeletons; the holes in the gaping floorboards resembled mouths full of bad teeth. Layers and layers of dust made me cough and sneeze. To begin with that old bothy was a hazardous place, but birds of all kinds favoured it as a nest site, including pied wagtails, wrens, swallows, a robin, a bluetit and several blackbirds. There had been so much clearing up to do on the farm that the bothy was one of the last places we tackled. Below the rotting upstairs loft space ferns had colonised the damp gaps in the cobbled floor, and water trickled and oozed down the back wall in slimy rivulets. We needed to sort the place out but it was low on the list of priorities.

The owl noises seemed to emanate from inside, but though I investigated thoroughly I could see nothing and went back to the vegetable patch. Then I heard it again. I didn't think of the chimney, because it had been sealed from the bottom and had a straggle of wire mesh across the top. The gentle hoots continued for the next few weeks but I could never quite work out where they were coming from.

Then one morning I rose early as a deluge battered off the Velux windows. As I was putting the kettle on, I caught sight of something resembling a soggy grey feather duster bouncing up and down by the door of the aviaries. It looked as if it was trying to get in. I had been brought an owlet a few days earlier and thought that by some extraordinary quirk it must have managed to escape from its box in the bird room in the house. I rushed in to check. It peered sleepily up at me from the box, clacked its bill in warning and then shut both eyes tight.

So the grey jack-in-the box leaping around in the rain was a wild owlet that had obviously been hatched somewhere in the surrounding woodland. It must have been caught out in the bad weather, and was now too drenched to get itself safely back to the trees. As I picked it up, I felt its tiny body under the sodden mass of down. I would have to take it in and try to dry it off. It was hungry and emitting constant food-begging calls. I poked a few pieces of chopped rabbit down it and left it with the other owlet in the box to recuperate. There was little difference in their size, though the owlet I had been given had sprouting feathers like little paint brushes beginning to appear. My intention was to put the mystery owlet back out later. Now it was raining so heavily that I did not want to leave it to get a further soaking.

Owls have long been thought of as being birds of wisdom and knowledge, but they have tiny brains, and I must now shatter the illusion and point out that as birds go, they are incredibly stupid. Compared to an owl, members of the corvid family – jackdaws, crows, rooks, magpies and ravens – would be well placed in the genius category. So I knew from sad experience that if I put this bedraggled baby out even under the cover of a leafy canopy, it would re-emerge to be saturated again. Many of the owlets that are picked up each summer are in a similar state. They leave their nests when at the 'brancher' stage and our frequent torrential summer downpours are lethal. After a heavy soaking, thick down becomes so sodden that they resemble sponges and quickly succumb to pneumonia and hypothermia.

The rain continued and later as I walked up the yard I noticed a grey, pineapple-shaped form on top of the old bothy chimney – an owlet. The swallows and house martins were anxiously dive-bombing and emitting shrill alarm calls. Unperturbed, it peered down at me with one eye shut.

This was obviously the sibling to the owlet I had found at dawn.

So the hooting must have been coming from the chimney, and the adults had succeeded in going in and out through the flimsy mesh cover. Later, as the sun finally broke through a sullen cloudbank, a third owlet emerged. All three gradually moved into the safety of the wood. Over the rest of the summer the tireless adult tawnies flew in and out to snatch food I put out for them. Excited hoots and shrieks from the adults, and constant begging squeaks from the owlets continued from dusk to dawn. The possibility that owls I had hand-reared had returned to breed so close was a moment for celebration. I could not have wished for a better outcome.

While the old bothy was being renovated to rescue it from total collapse, we put up several owl nest boxes around the farm. One nearest the house was swiftly adopted, and the following year our owls had another trio of owlets. However, once the bothy work was complete, they moved straight back to their favoured chimney. They have been using it intermittently ever since.

Some years there are sagas when the owlets get stuck on the roof and Iomhair has to get a ladder to climb up to rescue them. There are no trees adjacent to the building and the inexperienced owlets must reach the wood by crossing an open stretch of garden. On occasions they have been found in the middle of the drive, where they are extremely vulnerable. As happened in that first year, it's not always easy to know if they are in residence. Owls nest early in the season but usually by May, as the growing young demand more food, the adults are under immense pressure. When I'm sitting silently in the garden I may hear the heated irritation of scolding blackbirds, wrens and robins: something is about to happen. A rush of air from wing feathers fringed to aid silent flight – an owl effortlessly cruises in with another vole, and the old chimney embraces it as it descends, revealing a flash of feathery knickers. It must be hard to hide away through the day when the neighbours are aware of your presence and are vigilantly watching for threats from behind sylvan lace curtains. If you want to know where the local owls are, then you will find no finer guides than blackbirds.

In Scotland the lovely barn owl is at the northernmost limit of its range. It is not as hardy as the tawny, and because of this, barn owl numbers may crash during hard winters. Throughout the year, I usually only look after a few of these beautiful, shy owls, and it's always exciting when they are with us.

During the enforced lockdown of the 2020 coronavirus pandemic my close friend, retired wildlife crime officer Alan Stewart, rang one hot June day to tell me that there was a brood of barn owlets in trouble at a farm near Glenalmond and asked if I could help. For years the family there had been enjoying seeing the owls hunting over their farm. It was a worrying situation for them; they felt responsible. Due to the suitability of the traditional unimproved grassland and the way they farm, barn owls have been with them for generations.

'What about lockdown?' I asked. 'We aren't meant to move more than three miles.'

'It's a wildlife issue and these owlets are at risk. I think you should go and see if you can do anything to save them,' came the reply.

Since March Iomhair and I had been no further than our local Co-op in Aberfeldy. It felt strange to be venturing further afield – on a guilt trip, literally, for I knew we weren't supposed to leave home. But barn owls are precious and we could do this with no contact with humans and avoid any risk.

Barn owls thrive in areas of unimproved pastureland, where tussocky grasses provide a plentiful harvest of small rodents. They often choose nest sites in farm buildings, derelict houses or hollow trees but will readily use nest boxes.

On arrival, we found a precarious situation. The owls had chosen a site under the floorboards of the old farmhouse, between the joists. It's a favoured place in derelict buildings. The ceiling had collapsed, bringing the nesting area down. Three little owlets cowered in the corner of the room. The floorboards were rotten, and it was a precarious walk across a floor strewn with roof debris as well as a cache of dead voles. In amongst it lay three dead owlets.

Voles make up the most important prey item for owls, and as these small mammals are cyclical breeders, in years of plenty barn owls thrive and sometimes rear two broods. Tawnies are only usually single-brooded. This was a vole year, as I had noticed copious vole runs in our fields and there were excellent reports of large owl broods all around Scotland.

Barn owls take some time to lay a full clutch of eggs, but they start to incubate from the first egg they lay. This means that owlets may hatch out as much as a fortnight apart. The three dead owlets had not died as a result of the ceiling collapse; one looked to be only a few days old at the time of death and had been dead for a while. The other two were around a week old when they succumbed. There was a considerable size difference between the survivors – one was particularly small and weak.

We were in a quandary. Though the maternity wing had crashed down, we could tell that the adult birds were still around as some of the voles were fresh. The adults had been coming in through an open window. However, the roof was unsafe, and jackdaws and crows could get in too and take the vulnerable owlets to feed their own hungry broods. The small area of remaining ceiling looked at imminent risk of collapse. We could move the owlets, but where? A move might run the risk of the parents abandoning them, something that is common with barn owls in similar circumstances. I desperately wanted to make the right decision. Should we leave them or take them? It took us a long time to decide. If we took them and they all died, I would feel responsible. If we left them and they all died, that would be grim too. I didn't want to interfere.

After much deliberation, we returned home with a basket of white down-coated owlets – three dandelion clocks – and a bag full of dead voles.

There was indeed a very large size difference – 100 grams – between the largest and the smallest. The biggest was probably at least a week older. It had been eating all the pies, but we were surprised to find how feisty the littlest one proved as it hastily snatched food, turning its back to ensure the older one didn't nick it. One fat vole went down whole in a rush. For the first few days, I fed them every few hours, making sure the two younger owlets had plenty. It didn't take long for them to settle, and it was astonishing to witness their growth rates. Their sharp little goblin faces began to alter, transforming their newly feathered facial discs into the immaculate shape of an apple cut in half. The bird room filled with fluff and dust from emerging feather coverts. The dogs had fluff stuck to their inquisitive damp, black noses, and were constantly sneezing and pawing at their faces to remove it. It got up our noses and made us sneeze too, and it stuck to clothing and socks, and followed us around wherever we went.

Though the rearing process is the same for both species, there are several differences. Barn owls tend to cache food – something that the tawny does not do. The box was filled with bits of meat hidden away in corners. They are far more wasteful, and messier. They are also incredibly vocal when hungry or distressed. It's a noise like no other – a weird wheezing sound – not quite a hiss, but certainly nothing you would relate to owls if you didn't know. Little wonder barn owls have long been associated with ghouls and witchcraft. Ancient unmanaged graveyards festooned in ivy make magical places for owls to live, with their abundance of small mammals, and the excellent roosting and nesting sites in churches and bell towers. This led to reports of hauntings – terrifying sounds in the middle of the night, strange eerie forms seen against a backdrop of a silvery moon. People thought the noise was the uneasy spirits of the dead wailing in distress.

If the owlets survived, they could be safely returned to their farm once they were fully-fledged. For now, I was keeping the concerned family, and Alan Stewart, informed of their progress. Their fate absorbed us all.

Telephone calls with friends usually included the now familiar statement – 'I bet you are getting lots of writing done during lockdown.' It seemed that everyone was either clearing out their attics, spring cleaning and redecorating their houses, or working so hard in their gardens that they could have opened them to the public. Though we had been busy renovating a shed, we had run out of wood, and there would be no more deliveries. And I was certainly not doing much writing either because there seemed to be so much going on around me. The weather was more like that of the Mediterranean than Highland Perthshire. The wildflowers were particularly spectacular, and I was lured on long walks with the dogs and my camera. Otherwise, it was a constant round of animal feeding and cleaning. 'Don't worry, it will soon start raining, and you will get that book written, so you must take advantage before it does,' said Iomhair – but it didn't. And the sunshine helped us all to endure the stress of the pandemic and the barrage of negative news that engulfed us in misery.

Soon the owlets needed to move outside to the aviary, where it would be harder for me to monitor progress. Incredibly shy, they weaved back and forwards as they huddled together while emitting their peculiar wheezing sound. In a larger space, they were not going to take too kindly to my intrusions – a good thing, because they needed to be wild, but I needed to ensure they were each getting plenty of food. While having several owlets together minimises the chance of them imprinting, adversely it means that perhaps one is getting less, or more, food than it requires.

The weather had turned even hotter, with record temperatures in the south of England. I was feeding the tawny owls every night as usual by leaving pieces of meat out for them. Revolting remains were dropped around the garden, and we had to find them before the dogs did. I am ashamed to say they are very gluttonous and down bones and rotten rabbit in one gulp – with repercussions. Inside the aviary, trying to keep flies from buzzing around the owls' meat and laying their eggs on them was another challenge.

Iomhair made nest boxes in the aviaries suitable for squirrels or owls to use. After a few weeks the owlets had barely left the safety of the aviary's house, then one morning I noticed all three huddled up together inside one of the nest boxes. All I could see was the top of their heads shyly swaying back and forth with eyes half closed.

Now they had lost most of their white down, and only a few stray wisps remained around their faces. Their buff, tan and white plumage looked immaculate; they had transformed into birds of extraordinary beauty. Females tend to have dark spots on the sides of their breasts as well as an apricot blush to the top of their chests, while males are whiter on the breast. It's not a fool-proof way of sexing them, but this is generally the case. It looked as though we had two females and a male.

Back at their farm, the family had cleared out a stone outbuilding and built a release pen within. We would return the owls to stay inside for a few weeks until they were accustomed to the area – this is a successful way to 'hack' owls and birds of prey back to the wild. As with tawny owlets, sometimes as many as 70 per cent will die in their first year. Making them a pen to contain them before release and accustoming them to a regular feeding site would give them the best chance of survival.

The family were out in the driveway waiting for us when we took the owls back home. Still in restricted conditions of social distancing, they greeted us with glowing smiles and copious effusive thanks from afar.

Though the two younger members of the family, Rowan and Laurie, had often seen the owls flitting around their farmyard, they had never been so close. Their mother peeped into the box, and gasped – I could see the tears in her eyes. I had a lump in my throat too – they had put so much work into clearing out the building and making it a perfect place for an interim owl refuge. With its high wooden beams and eaves where they could perch, and plenty of space for flight practice, it was ideal.

As Rowan and Laurie held the owlets, clutching them tight in the hold I had shown them, their excited faces said more than any words. Here were two more young people signed up to the importance of nurturing the wild – the connection had been made. I felt overwhelmed, too, particularly when they gave me a bunch of flowers and a beautiful card as a thank you.

Dappled sunlight and a tangerine glow filled the evening as we drove back home through the glen. Though spring at home had been spectacular with its rafts of bluebells, dawn chorus and clear azure skies, we had missed spring everywhere else. We stopped to watch lapwings and oystercatchers with chicks beside a bank of field pansies and red campion backlit in the low rays of the evening sun. Mournful cries of curlews drifted from far out on the hill. Nature is perfection. Thanks to Alan Stewart and the family at Glenalmond, three barn owls had a chance to eventually rear broods of their own. It was a team effort – one that brought out the best in human nature.

20

Helen and Pipkin

Now that the habitat in the surrounding area has improved to such a degree, and the local squirrel population is thriving, I sometimes take in young squirrels that have been hand-reared by other people for release here. It works well, and nearly all the ones that have been in the aviary for at least a month prior to being set free return to feed in the garden. Most of these squirrels come through the Scottish SPCA, but I also occasionally receive various mammals and birds from Hessilhead Wildlife Rescue Trust.

When it comes to wildlife, it would be hard to find two more dedicated human beings than husband and wife team Gay and Andy Christie from Beith in Ayrshire. I first came across them at a wildlife rehabilitation conference in a hotel in Inverness some thirty years ago, and have found their friendship, help and advice invaluable ever since. That first time I met them, I remember being amused that Gay had to keep dashing out of the various talks to return to their bedroom to feed a litter of tiny brown fox cubs. They were cosily snuggled up together in an electric brooder. It highlights my point that wildlife never appears when it's convenient and you have to work around the situation with which you are faced. Who'd have thought of a litter of fox cubs in a hotel bedroom?

The Christies started their venture nearly fifty years ago, when they were first married and living in a council house at Lochwinnoch. Andy took a fox cub from a gamekeeper and they reared it in their home. It was not long before the council house premises began to overflow with various injured creatures from ducks to swans, geese, hedgehogs, owls and hundreds of small birds. Nothing was ever turned away. Gay and Andy knew that this was the work they wanted to do for the rest of their lives, but they needed to move to bigger premises and find a way to fund their venture.

Today Hessilhead has become Scotland's largest independent wildlife hospital. Extraordinarily well hidden in a wooded area of deepest Ayrshire, during the time since they acquired it they have expanded a 20-acre site into an astonishing rehabilitation facility that includes an intensive care unit, a swan and seal unit, a surgery, sixty outdoor aviaries and pens, and an impressive hedgehog building. They also have a purpose-built classroom where students, vet nurses and members of the public come on courses to learn more about wildlife rehabilitation and first aid.

Like me, they are despondent about the current state of the natural world, our lack of respect for it, and about the role we as humans are playing in the demise of a great many common species, but also in some of the harrowing incidents that befall wildlife due to development and urbanisation. Like me, too, they have seen numbers of starlings and sparrows plummet, and now see fewer members of the tit family, and very few kestrels. As I've mentioned, hedgehog numbers continue to fall dramatically, though you might not believe this if you could see the huge influx they receive at the end of summer. Song thrushes were once regular patients but are now a rarity – are toxic slug-pellets another contributing factor in their demise?

Fishing line and hooks as well as collisions with wires and traffic remain a serious problem for swans and other water birds. Gulls frequently fall foul of our detritus – they get caught in the plastic rings that hold six-pack drink cans with disastrous results, and increasingly ingest plastics as they feed at sea. Now they are getting caught up in the mountain of irresponsibly discarded facemasks associated with the coronavirus pandemic.

Rehabilitators often witness how people expect wildlife to be there for what they want, feeling it's all right if it's conforming to their notions of 'good' behaviour, but when it doesn't conform, or gets in the way, it's a different story. Foxes are a prime example – councils and housing departments pass the buck, or the fox in many cases, to people like the Christies, and the Scottish SPCA are called in when someone suddenly decides they don't want a fox in their garden. Some people still appear to think they eat grandmothers and children as well as cats and dogs. Where we live we seldom see foxes, though there are plenty around. In the urban environment the modern fox about town is far more partial to a chicken takeaway than a fresh one from a henhouse, and often takes advantage of the messy packaging and waste food that we chuck everywhere.

As well as hundreds of birds, deer, badgers, otters, pine martens, red squirrels, seals and bats are all part of the day-to-day life of Hessilhead. Unusual patients have included a storm petrel (our tiniest, sparrow-sized seabird) found on the doorstep of a Sauchiehall Street nightclub at midnight, and a pomarine skua found in Strathclyde Park. I too once received a storm petrel one dark winter's night, when a policeman who I knew well came knocking on the door. When I looked out of the window and saw him standing there, I feared the worst, my heart raced as I went downstairs to let him in. He gingerly passed a tiny black heap of feathers into my hand and said, 'I have just found this in the middle of Crieff and thought you might be able to save it.' I was astonished when I realised it was a storm petrel. Crieff, in the heart of Scotland, is about as far from the sea as it's possible to be in this country.

Birds caught in discarded cooking and sump oil are frequent, and the use of inhumane 'glue boards' intended for rodents and still readily available, trap unintended victims including birds and hedgehogs, who are left helplessly stuck to die a prolonged and agonising death. I once received a helpless red squirrel kit that had met the same glutinous fate – happily it was one of the lucky ones and though it took several attempts to remove the glue residue from its feet, I was eventually able to let it go again. We have a lot to answer for.

One day, the Christies brought me a young female squirrel for release, and after we had put her safely in the aviary and were sitting chatting over tea, not expecting much of a response, I asked Gay, 'Has the squirrel got any specific dietary requirements?' She replied immediately, 'Yes, she loves nectarines.' So the new baby was instantly christened Nectarine, and I had to go shopping. Squirrels always intrigue me with their favoured foods and some of them are very picky, discarding many of the things that are relished by others. Even after release Nectarine returned regularly to partake of a slice or two of this delicacy from a dish on the aviary roof, and sometimes I noticed the resident squirrels trying it out too. Not all of them were impressed.

*

Though I prefer not to name the squirrels that pass through each year, there are the inevitable exceptions. Some might be temporarily named after the people who bring them. Helen was one such. She appeared in the summer of 2020 during the lockdown phase of the coronavirus pandemic and both her finder, my friend ecologist Helen McCallin, and I drove further than we were meant to in order for me to collect the little waif. Squirrels are important.

While out walking through woodland paths near Forfar, Helen's aunt and uncle had found the tiny squirrel. She had suddenly appeared and immediately started trying to run up their legs. She was obviously starving. Though it's not unheard-of for a young squirrel to become emboldened when desperate, it's not normal squirrel behaviour. Close by the path the couple also found a small wooden box with hay in it left lying on the ground amongst the ferns. It pointed to the fact that someone had hand-reared this baby and then let her go. But why had she been released far too soon? What had made them push her out so early? Had the squirrel simply become too much of a handful? It was – and remains – a mystery.

When I met Helen in a car park in Dunkeld she had the tiny squirrel in a box on the passenger seat. A little face peeped out of fleecy bedding and I saw instantly that she was indeed very young. The little squirrel still needed milk. She was at the awkward phase of not yet being fully weaned. Once home she took a syringe of milk without any battle – unusual in a wild baby, confirming our suspicions that she had been hand-reared. It did not take very long before she was sampling all the solid food I offered her. Some squirrels are far messier than others. By morning, the floor of Helen's box was littered with half-chewed pieces of food, and much of it was taken into her bed. Her fleece hat nest was filled with cached food. It must have been most uncomfortable; I thought of the occasions when I have eaten toast in bed, and then regretted it because the inevitable crumbs have the power to annoy you however hard you try to clear them off the bottom sheet.

Helen McCallin and I continued to wonder about Helen's past. We would have loved to have spoken to the person who had first rescued her. Other than being starving hungry, she was a fit little squirrel with a glossy red coat. Eventually I was able to re-release her. I don't suppose I will ever find out the full story, but someone had done a good job. Their only mistake was that they set her free far too soon.

*

In 2019, a month prior to Christmas, I took in a squirrel from one of the girls at the Scottish SPCA. She was a late baby and had been reared on her own. She was still too young to be released, and the idea was that she should be overwintered in the aviary and then released in spring. However, that was not to be.

She settled well and I kept my distance, even though I knew she was going to be with us for a long time. I had seen the way she had behaved with the person who reared her, racing all over her, and lying on her back in her hands to be tickled. Now we had to back off. I took care not to go in to feed her when she was out of her sleeping box. This seemed to be working well, and I simply kept my eye on her progress from a distance.

However, just before Christmas I was cleaning out the aviary and I noticed that the droppings on top of the feed table and the sleeping box were not normal. When you are working with animals, or are out looking for wildlife, you spend a lot of time studying the minutiae of muck, or scat, as it is also known – it can tell you not only what is around, and what that particular animal or bird has been eating, but it can also reveal a great deal about the health of the animal. A very serious American woman who I was taking out on a wildlife tour once asked me if I had a degree. I replied that yes, I had a degree in Turdology. Say it fast enough and with the intonation in the right place it sounds impressive. My joke flew over the top of her head and all she said, was 'Ah yes'.

She didn't enquire further. Knowing about 'turds' is vital if you want to understand animals. And if there is blood in the excrement, then action must be swiftly taken. I noticed that not only did the squirrel have diarrhoea, but there were faint bloody traces in it too. As I had not been in for a couple of days I was fearful that perhaps I was now too late to do anything about it. There was no sign of the squirrel, so I nervously investigated the sleeping box, hoping against everything that she had not already died.

She seemed very pleased to see me and despite the distance I had been keeping between us, leapt out of the bedding and climbed into my hands. I would have to take her into a hospital box in the house and hope we could overcome the problem.

Though lethargic, she had not lost her appetite. This was a very good sign. Wildlife vet Romain Pizzi prescribed paediatric oral antibiotics and quickly organised a prescription. Unfortunately, this meant that I would have to handle the patient twice a day. I was none too keen on the idea, particularly when I was trying desperately to make her less familiar with humans.

There followed ten days of twice-daily handling to give my patient banana-flavoured medicine; she took it well to start off with but it became increasingly difficult as she quickly realised my intrusion was not going to be pleasant. I was fortunate not to be bitten and had to work swiftly to get the job done. Once the medicine session was over, she launched herself towards me whenever I went to feed or clean her out, or take in more food. She was simply the tamest squirrel I have ever had in my care. She needed some exercise to burn off the high-octane energy so typical of squirrels. Shutting all doors and windows tight I locked myself away with her so that she could race around the hospital room. She quickly discovered that I was the most similar thing to a tree she could find and as I stood motionless she used me as a racetrack. From the top of my head to my shoes and everywhere in between she flew randomly round and round, squeaking gleefully, and if I stretched out my arms she would run down them too and leap onto the other box cages in the room, before racing up and down my back, legs and shoulders over and over again. Iomhair came in and had the same treatment, but he was less enthusiastic when she leapt onto his head and scratched the thinning areas. She was fearless and even if it had been spring and time for her release, this overly tame state meant she could not yet be released. More tough love was needed but I had a nagging thought in my mind that this was an exceptional squirrel, one that might not return to the wild. Time would tell.

Christmas passed and I learned that squirrels are also very partial to Brussels sprouts. And every day I found out what it must feel like to be a tree, climbed on and used as an adventure playground. Such speed, agility and precisely measured leaps and loops; all at dizzying speed.

Many months of cold, grey rain lay ahead. It was going to be lonely for a squirrel kit with no interaction with either other squirrels or me. Once she returned to the aviary, I gave in and let her play with me, and hoped that by spring, another young squirrel might be brought in to keep her company and eventually help her to revert to squirreldom instead of a world dominated by humans. Then I hoped that following a period of social distancing from me the two could be released together.

Every day she launched herself towards me chattering and chortling as she climbed into the hood of my jacket, or investigated my pockets. Her sharp little claws pulled threads all over my jeans, and she nibbled my watchstrap and a new belt before I even noticed. Sometimes she wanted to lie on her back as I had seen her do with the person who reared her, so that she could be stroked. Then she made contented little squeaks and wriggled with delight. How was she ever going to return to the wild being like this? I had now named her Pipkin. Pipkin was imprinted. Pipkin was a challenge.

The Scottish SPCA had numerous young squirrels at the start of the year and after I explained the situation, they helpfully brought a single female to me to see if putting her in with Pipkin would help her revert to wildness. This new squirrel was certainly terrified and did not tolerate humans in any way. She swiftly belted into her nest box as soon as I appeared. She was going to have no problem with going back into the wild. I stepped back again and made sure I did not go in when Pipkin was awake. Though the two squirrels were together for nearly two months, it made absolutely no difference to Pipkin's behaviour. She was still far more interested in associating with humans than squirrels. The new youngster was duly released into the garden.

I agonised over Pipkin, but it was simply unsafe to release a fearless squirrel. It would be guaranteed to end in disaster – and very quickly. Then during the autumn another late-born female baby squirrel arrived – it was going to have to overwinter with us as it was now too advanced in the year to release such a small youngster. Bizarrely, within weeks of the baby's arrival Pipkin began to ignore me. She stopped using me as a trampoline and instead the two spent hours playing together. They seemed to have an extraordinary rapport, and also slept together in the same nest box. Pipkin became more and more aloof. In fact, she suddenly seemed to revert to wildness. This was a real breakthrough, but sadly it was now far too late in the year and the weather far too erratic for successful release. So the two female squirrels will overwinter together. Up until this point I felt I had failed Pipkin. When you work with wildlife there will inevitably be animals that slip through and cannot be returned to the wild, like Sage, that first tawny owlet, reared thirty-three years ago now, who also became imprinted on me. I had consoled myself that Pipkin had never known freedom, but it looked as though she was starting to learn to stay away from humans. For now, as I write this in 2020, I will keep my distance and hope for a glorious spring when both squirrels will gain their freedom. Not all decisions with wild animals are easy. Pipkin has challenged my ethos of always letting go, but with any luck that may still be possible.

21

Cloudy

It has been extraordinary to see that in these recent troubled times, when our lives have been altered out of all recognition, more and more people have turned to nature to salve the misery, angst and uncertainty that has affected each and every one of us. To begin with the so-called lockdown forced us to stop rushing about like aggravated wasps attracted to a sticky jam jar. Roads and air space went quiet, birdsong could be heard above the usual deafening racket of our frenetic lives, and in the ensuing peace people became aware of a world they had never experienced before. And by way of respite they began to watch many of the little things that surround us – the minutiae of a previously unseen, and definitely unappreciated, world. Even in the urban environment the stirrings of nature – the previously overlooked lives of garden birds, mammals, insects and flowers – kept us focused and enchanted as never before. Surely this is one positive aspect to emerge from the gloom? My hope is that those who have discovered solace, and perhaps were less aware of the power of nature beforehand, will retain their love, and if and when we are able to move and mingle as we did previously, they will remember the part that nature plays in salving fears and concerns. They will hopefully understand too that nature is the key to our future success on this planet. If only we could feel confident that this will be the outcome. Sadly, I remain riven with doubts and fearful that many will simply revert back to their old ways.

I think by now you will have understood that animals have always been at the heart of my world. Since my earliest childhood I have had the desire to know and understand their needs, and recognised the basic fact that whatever we may think, we are not superior beings. Animals are not conveniences to be picked up and dropped at will. We abuse, torture and cull them, and often keep them in dire conditions. Instead, we should be holding them in high esteem, for each is a sentient being that needs our respect and our compassion.

During the spring and summer of 2020 – during the time I was writing this book – we received far more orphaned and injured birds and mammals than usual. Probably due to the pandemic and enforced lockdown, when we we were only allowed out for a walk locally every day, more people who would not usually have been out in their local area were finding creatures in trouble. Though we have experienced the inevitable losses and concerns associated with working with wildlife, overall the arrival of so many creatures that needed care, as well as the intrigue of the lives of the wild things in the surrounding area, and the finest spring and summer for decades, brought us a welcome diversion. It also brought joy.

*

It is 1 June 2020 and high pressure lies over Scotland, bringing eternal cornflower-blue skies and sweltering heat. I am walking our three generations of collies – Molly, Maisie and Pippin – early, before the day becomes too hot. Above the farm, broom is showing off. It glows a vibrant, electrifying yellow. Bird cherry and hawthorn blossom add to a melange of intoxicating scents as swallows hawk for insects over the pond. They skim low over the water as sunlight paints their blue-black feathers and enhances their crimson faces – little fragile china dolls with rouge on their cheeks. The songs of blackcap, chiffchaff, blackbird and wren drift through the flowering scrub. It's 8.30 a.m. and heat is rising, the valley beneath us shimmering with the promise of another burning day. Pippin is fifteen this year and she is slowing; she stops at a small burn trickling down through a tangle of birch trees and lies in a pool to cool her tummy. An orange-tip butterfly drifts slowly past as we start to climb the steep hillside still studded with late bluebells and the delicate little faces of dog violets. There's something lying on the path in front of us. Hunched like a discarded old tweed coat, it lies motionless while flies buzz and drone in a halo around its prostrate form.

I stop and make the dogs sit behind a large rock while I creep over to have a closer look. It's right on the path out in the open, away from all cover – a path frequented by an increasing number of walkers and their dogs. It's lying face down and I can see that its dappled coat is sticking up like a tent along the bony ridge of its spine. It's a badly dehydrated deer calf, perhaps five days old at most. I creep closer, signalling to the watchful dogs that they must not move – miraculously, they don't, but quiver with excitement, ears pricked with intrigue as they watch every footstep, eager not to miss out.

Then I remember that last night as we sat in the garden watching the bats emerge from the old bothy, we heard impassioned shouting coming from this area. Someone was roaring at dogs. The yelling became desperate, and a volley of angry expletives shattered the stillness. We hadn't thought much of it other than that someone had lost control.

When people ring me about the newborn deer they have found – most often it is roe fawns – unless they have found the mother dead, I usually suggest that they back away quietly and don't disturb them. However, here something is definitely not right. Not only is the tented skin of concern, but the calf is also crawling with ticks and flies. Perhaps the hind was frightened away. Perhaps all the shouting we heard was related to a dog chasing her away. It is a coincidence to find this calf here, and now. It seems as if it has not fed for some time, and that's unusual. Bluebottles continue to congregate – the semi-comatose calf is an ideal site on which to lay their eggs. I think of the resulting maggots, how they will literally start to eat the calf alive, squirming their way into flesh. I ring Iomhair as I don't think I can manage to carry the calf back down the steep hillside while also controlling three curious collies.

I sit on a rock to wait, my feet on a carpet of celandines. All around me ferns are beginning to open, and the bracken too is starting to emerge. Then I see him hurrying up the sheer hill. Breathless and puce in the face with the exertion of the climb, he takes a moment before he speaks. 'I wouldn't leave it there,' he says firmly as he looks over at the calf. He knows well how I enjoy having young animals to look after too. We stand and deliberate further – we both want to do the right thing. The ground where the vulnerable calf lies is so open. It's not the usual kind of place for a hind to leave her offspring. 'I definitely think we should take it,' Iomhair says. I put the dogs on their leads as they are effervescently over-eager to assist, and we set off back down the hillside. We make an odd cavalcade with Iomhair carrying the calf wrapped in a towel. Its head sticks out. Momentarily I look and see how similar it is to that of a wallaby, yet the two species are not related. The dogs tow me down the bank at high speed. Twice I trip over them as they jink in front of me tying my legs with the leads as if we are in a three-legged race. I curse and growl but they are far too intrigued by Iomhair's mystery baggage to take any heed of my exasperation – they are only being dogs. By the time we reach the farm we are both flustered and red in the face. The dogs collapse in the burn lapping and wallowing before shaking beads of cool water all over us.

Once safely in a pen in the shed, I see that the calf is a hind. As I thought, she is severely dehydrated so something must have happened to her mother, and we reassure one another that we therefore have no regrets about taking her. I give her water, glucose and electrolytes to rehydrate her. We estimate she is perhaps a week old, perhaps even less, and I hope this indicates that she will have had the all-important colostrum from her mother before they were separated. She is quiet and calm – in shock – and yes, she is covered with tiny ticks, but few are attached or yet bloated with her blood.

By lunchtime she half-heartedly takes more water and a weak solution of milk replacement but she is still not suckling. It will take time for her to accept the teat. I will have to be patient until she gets used to the bottle and the alien rubbery nipple. After three days she is suckling well, and after a week she is making excellent progress. All too aware of the things that can go wrong we refuse to give her a name yet as we don't want to tempt fate. Just as I had been with Ruby, I am already besotted, but unlike Ruby this calf is far fitter, and she is not affected by tick paralysis. I wonder if this will make our journey together easier. Meanwhile, outside in the field Ruby peers in through the shed window and calls to me constantly to come out and use the hosepipe to give her a cooling shower. We have spent years trying to fix the boggy areas in our fields and repair the old blocked field drains, and now a wet, muddy wallow would be much appreciated. Deer love to wallow, and Ruby loves water and will stand patiently for a long time while I hose her. And when I clean out the field water trough she gallops down, and if it's a warm day, stands as I pour buckets of water over her back and neck. However, she won't let anyone else do it and I always end up getting soaked in the process too.

The new calf dominates my days – five feeds, starting at 6 a.m. In between it is a continuous round of animal feeding with barn owlets, hedgehogs and baby squirrels to look after too. I smile when I reflect on the frequent comment, 'You must be finding a lot more time to write now we are in lockdown and you won't be out and about so much.' The fine weather continues day after day, and after nurturing the growing flock, an old-fashioned and now rare wildflower meadow adjacent to the farm lures me like a strong drug. I cannot keep away.

*

The State of Nature Report 2019 does not make for comfortable reading – the abundance and distribution of Scotland's species have continued to fall in the past decade, with no let-up in the net loss of our precious biodiversity. These declines include an estimated crash of 51 per cent in pollinators, including bees, hoverflies and butterflies, since the 1980s. As well as being glorious additions to our diverse fauna, they are also vital to our survival – 84 per cent of our crops depend on pollination. Pollinators are essential to the success of our economy and are crucial to food production, and therefore to our health and wellbeing. The famous saying (often attributed to Albert Einstein and other renowned commentators) that 'If the bee disappeared off the face of the Earth, man would only have four years left to survive' is a stark warning, but have we taken note?

Numerous factors have led to this dire situation: the loss of 95 per cent of wildflower meadows and associated habitat degradation. Then there is the relentless use of pesticides and agro-chemicals in intensive farming systems, a huge expansion of building and urban infrastructure and subsequent loss of woodland and green spaces. We are over-zealous gardeners and carry out unnecessary strimming in our obsession for tidiness, while the cutting of grass verges beside our busy roads is also detrimental. Many of these verges do not impede sightlines or cause dangers for traffic yet are still cut. These changes have led to the demise or isolation of insect populations, marooned like the red squirrel in small pockets of suitable habitat where they struggle to survive. Every existing wildflower meadow is important, and like the establishment of new native woodland, can provide a rich habitat for a wealth of species.

In 2017, the Scottish Government approved a 'pollinator strategy' to help protect, enhance and create flower-rich habitats both in the countryside and the urban environment. Since then, numerous inspirational and optimistic projects have taken flight. Though this small meadow close to our farm is set close to a sea of intensive farming on the Tay Valley floor, and is but a tiny remnant of a bygone farming era, it highlights the importance of linking mixed habitats to create safe corridors for all our wildlife.

In spring bluebells, stitchwort and wood anemones spill forth from the woodland edge above and fringe the meadow, amid intoxicating scents of blackthorn, hawthorn and wild cherry blossom. The squirrels are gathering moss and lichen off the old drystone dyke that runs up the hazel wood. They bounce cheerily along the grizzled stones with overflowing mouthfuls of soft material for upholstering their dreys. Blackbirds and thrushes already have youngsters and I can hear their continual 'chirrups' from a thicket. Warblers and other summer migrants have been continuing to arrive, adding new melodies to the avian dawn concertos.

The meadow is dynamic. Every day brings a new discovery as the bare, winter-grazed sward recovers with surprising speed. Buttercup, red campion, ragged robin, crosswort, bedstraw, lady's mantle, bird's foot trefoil, orange hawkweed (nicknamed fox and cubs), self heal, red and white clover, cranesbill, eyebright, ox-eye daisy, bugle, yarrow and cuckoo flower, are just a few of the species that find their place in this pastoral paradise amid vital meadow grasses including cocksfoot, crested dog's tail, quaking grass, sweet vernal, Yorkshire fog, bromes, false oat, and a selection of plantains and sorrels. And then there is yellow rattle, a plant that parasitises some of the more dominant grass species and helps to keep the sward open for its weaker floral companions. And usually around mid-June the meadow is lavishly sprinkled with orchids – some years they cling low to the ground, but when it has been wet, they grow tall, their delicate pinky-mauve and white or purple heads peeping above waves of feathered grasses waltzing in the breeze.

I cannot tear myself away from this place amid its continual thrum and the busyness of insects, from the tiniest iridescent green spiders and shield bugs to the flamboyance of new butterflies and day-flying moths – fritillaries, commas, peacocks, meadow browns, common blues, ringlets, orange tips, small heaths, and if I am lucky I may find a male emperor moth, one of the most beautifully dressed moths of all and occasionally seen at this time of year, attracted by the strong aroma of the female emperor's enticing pheromones. Grasshoppers tick and whirr as sunbeams light a froth of cuckoo spit created by a minute froghopper. Tiny dark chimney-sweeper moths rise from a floral mat. At dawn dew beads the nodding heads of thousands of grasses, and for a short time heavenly birdsong shuts out a world of worry and I am absorbed in the microcosm of hundreds of invertebrates that are far more important than us. A brown hare leveret lollops lazily over the rise and stops to nibble succulent new clover leaves. In among all these treasures are dozens of bees, hoverflies and a wealth of other insects I struggle to identify. And every year I forget more than I have ever learned. It's all part of being intoxicated by nature.

I rise early each day at dawn to feed my new charge and then lose myself for an hour or two in the meadow, where on every visit I find something new. The calf is doing well and – now feeling a little more confident that she is a survivor – I name her Cloudberry, Cloudy. Like Ruby, she quickly answers to her name and calls to me when I go in each morning. I am mastering the fine art of cervine mewing and as she responds, I think I must be improving my hind contact call. There is always room for improvement.

Cloudy accompanies me around our fields for walks every day and we venture into Kim's Wood together to collect her armfuls of fresh browse – the leaves that deer love best: aspen, cherry, hazel, field maple, rowan, hawthorn and wild crab apple. She loves raspberry and willowherb too and during the night eats her way through an astonishing bouquet of succulent greenery. By morning the stems and branches are stripped bare. Ten years ago, before we added deer fencing around the perimeter of the farm, we made a small paddock in the corner of a field and had deer fencing put up there especially for Ruby. Then it was simply a grassy bank but now, having been relieved of grazing pressure, nature has taken over the area and hazel, hawthorn, cherry, oak, alder and rowan have sprung forth. There is plenty of cover for Cloudy, and here she can prune to her heart's content. With long grasses and herb-rich wildflowers, she spends her days contentedly waiting for me to bring her the next bottle. She trots after me like a dog, but then when I take her into our big fields she takes off randomly, leaping and racing with the speed of a gazelle, vanishing into the flowering grasses before racing back to me and showing off with high kicks as she jinks madly in all directions.

There is no doubt that we will look back gloomily on 2020, the year that will long be remembered as a 'non' year – everything cancelled, people working from home or losing their jobs, shops and tourist venues in crisis, no parties, no celebrations, no book festivals, theatre or concerts, all obliterated. And no physical contact. Standing outside our local Co-op speaking to a dear elderly friend who has just lost her husband after fifty years of marriage, I cannot even put my arms around her and give her a hug. Instead, we stand at a distance and wipe away the tears and I come home feeling the intense cruelty of the situation – as well as totally useless.

So Cloudy's arrival, exactly ten years after we received Ruby, is a welcome diversion, and I feel incredibly lucky for that. Unlike Ruby, Cloudy has not had the health issues to overcome and though I know that things can always go wrong, I am hoping that she will be with us for many years to come. Introducing her to the sheep and Ruby will be the next step. Though they chat through the fences, I fear that Ruby, who is definitely the matriarch of the flock, and has the size and weight of a fat donkey, will probably give her a hard time. First I will put Cloudy in with some of our more gentle sheep.

Though for the first few days after they are born, deer calves are left by the hind and lie patiently awaiting her return, soon after they follow her everywhere and can run every bit as swiftly. Together a hind and her calf may tirelessly cover vast distances over very rough ground during a single day. At this stage Ruby did not much like being left alone either, though once I was out of sight she would relax again. As I had done with Ruby when I was writing *Fauna Scotica,* I set up a small desk in the doorway of the shed and wrote much of this book with one small red deer calf in attendance. As I work she potters off into the paddock to graze and then trots back and forth and lies in luxury on a moor-it-coloured Shetland sheep fleece beneath my makeshift desk. The concrete floor is cold! She tucks herself in at my feet and cuds peacefully as I type. In the trees in front of the doorway as I look up or am pausing for thought, I see squirrels dashing about, always in a hurry, always busy. One scampers past us, oblivious of our presence. Robins come in through the open shed doorway and perch cheekily on the wall. They peer expectantly down at me, heads set quizzically, knowing that I will scatter biscuit crumbs for them.

22

Squirrels on the Move

Many landowners have been working towards reinstating suitable habitat for squirrels, with successful planting schemes of a mixture of broadleaf and coniferous forest enabling the return of squirrels to areas where they have been absent in living memory. Due to the isolation of many of these new woodlands and the lack of habitat connectivity between them and areas where squirrels currently thrive, these small, vulnerable mammals are never going to recolonise without assistance. Translocation has become a highly successful conservation tool, and ornithologist and wildlife consultant Roy Dennis has been at the forefront of this valuable work.

Roy Dennis's name is usually linked with ospreys, for after he met the late George Waterston of the RSPB, who instigated setting up a public viewing hide at Loch Garten, on Speyside, he started his lifelong work with the species. Waterston's brainwave altered public perceptions of wildlife, with 14,000 visitors viewing the ospreys in 1959. Roy worked there as warden from 1960 to 1963 before moving to Fair Isle. By 2001, visitor numbers to RSPB Loch Garten had reached an astonishing 2 million; enthralled by what they could see there, it was clear that people were beginning to care deeply about the wellbeing of the ospreys, and had become acutely aware of the threats of continual persecution. A combination of enlightenment and Roy's dedication, aided by that of a host of other enthusiastic raptor workers, has without doubt led to their recovery. There are now well over 250 breeding pairs in Scotland. The osprey's recovery is one of the most successful British conservation stories of all time.

A passionate advocate for the reintroduction of lost species, in particular beaver and lynx, Roy has been involved with the reintroduction of white-tailed sea eagle and red kite, as well as golden eagle to Ireland. His pioneering satellite tracking work over the past twenty years has not only provided valuable scientific data, but it has also spawned highly popular web-based public access. Being able to watch the birds' progress over the internet allows people to follow the fortunes of globally migrating raptors, proving the importance of public participation in wildlife conservation matters.

In 1995 Roy set up the Highland Foundation for Wildlife – a non-membership charity dedicated to conservation and research, with the focus on species recovery ventures and habitat restoration. Based near Forres, Roy spends his time lecturing, writing and travelling to advise on and oversee environmental projects all over the UK and abroad. One of his many success stories, and one that brings hope in a world where there is so much negativity concerning wildlife, is his red squirrel translocation work.

When Jane Rice, the owner of Dundonnell Estate in Wester Ross, contacted him in 2005 about the possibilities of bringing back red squirrels to an area she saw as the perfect habitat for them, work began on surveying to check its suitability. The last records of squirrels here were from the 1970s, but due to loss of suitable habitat, they had died out. He told me, 'Many woods had been felled and replanted, and though young forestry is good for such mammals as pine martens, it is no use for red squirrels until it reaches at least twenty years old. It's vital to be sure the area is going to provide all they need, and Dundonnell Estate woodland seemed to have been transformed and become suitable for squirrels. I began to look into moving some from areas where they were thriving in Moray and Strathspey, around Dunphail, Edinkillie, Carrbridge and Boat of Garten. Following communications with the then Scottish Natural Heritage – now NatureScot – I was given a licence to trap squirrels in places where they were coming to garden feeders. This work involved me in numerous visits to ensure that the population continued to grow unaffected by removing a few individuals.'

To translocate squirrels, live traps are set close to the squirrel box feeders and first left open baited with nuts. The squirrels quickly become used to them, and then the traps can be simultaneously set. It's a laborious process as not only do squirrels have to have come from different areas to avoid close genetics, but before a change in the law, every trapped squirrel had to be taken to the Strathspey Veterinary Centre at Grantown on Spey. Here a vet anaesthetised them and weighed and measured them, taking blood samples, checking them for parasites and fitting them with microchips. 'We had to do this with the early translocations, but subsequently they could go straight away to the release sites; it's pretty obvious from their appearance when squirrels are fit and well. However, these were the rules, and as soon as they were deemed fit – which all have proved to be – they were put into individual nest boxes lined with hay and containing nuts and sweet apple (the latter avoids dehydration on their journey). They were then taken straight to their release location.'

In early 2009 the first squirrels were taken on the two-hour car journey to Dundonnell.

'On arrival, the boxes were then quickly fitted to the trees above nut feeders and the doors taken off. We then plugged the entrances with dried grasses or moss so that they remained shut for a while and then the squirrels could push them open later when we had gone.

'Initially, there were suggestions that the squirrels should spend the first few days in a pen before a soft release, but this doesn't work. Sometimes two squirrels may not get on with one another, and being in a pen is stressful, as they have to remain in close proximity. Out of a total of forty-three caught that particular winter, we only lost one, and that was in a soft release situation. It proved that it is far preferable to put them straight out. Dundonnell keeper, Alasdair MacDonald, was a vital part of this particular project and kept detailed records. He also ensured that feeders were always topped up. During that first year, he used eighteen sacks of peanuts, twelve sacks of sunflower seed and six of maize.

'The first signs of pregnant squirrels were seen in May 2009 and by that winter, when the leaves were off the trees, twenty-six dreys were located and eleven squirrels counted in just three hours. Keeping tabs on numbers is difficult due to the habitat. We fitted four squirrels with radio tags; two of the females had two litters each that season. Like many young males, one clearly had the wanderlust and crossed the mountains to Loch Broom. He remained in one big garden, and in March a female was released there, and they bred. It's the first time I have played matchmaker to squirrels. Eventually, when it was clear that the population at Dundonnell was burgeoning, twenty were caught in 2012 and moved to Loch Broom. Now there are red squirrels spread out over a large tract of woodlands between the gardens of Ullapool and Braemore Junction, and at Dundonnell. It's an excellent result.'

Similarly at Alladale Wilderness Reserve and neighbouring Amat and Croik estates in Sutherland, thirty-six squirrels were taken from Moray and Strathspey in the early part of 2013. They, too, are thriving and bred in their first summer. When

Roy checked on one of his monitoring visits in November 2015, he counted eleven in one garden, and it was reported that some had spread 10 kilometres along Strathcarron, and now they have spread into the Strath Oykel forests. A local farmer also witnessed a squirrel swimming across the River Carron. 'If the population is expanding, individuals will begin to cross over mountainous areas and find ways to cross rivers. Then they will start to spread out. In my opinion, where forestry work entails large tracts of clear-fell it would be relatively simple for rangers to move squirrels in this same manner. We should also start releasing reds in areas where the population of grey squirrels has been reduced; when people see red squirrels, it really helps to change their attitude and would encourage them to be more proactive in humanely ridding the countryside of the grey.' The need to remove grey squirrels often leads to misunderstanding, concern and conflict, and it is indeed hard to accept without understanding the context of their impact on the far more vulnerable reds.

Roy and his colleagues moved their translocation projects to suitable new sites in the north and west of Scotland. This valuable work is heavily reliant on funding and cooperation, as well as the dedication of landowners such as Jane Rice and Paul Lister, the custodian of Alladale. They now bear witness to its enormous success. It's an investment with priceless dividends – squirrels beginning to reclaim old haunts right across the northwest Highlands.

Roy Dennis is a man who has always made things happen. He is adamant that there is sometimes far too much scientific study and time spent surveying, and not enough getting on with it. He was awarded an MBE for his work in Scottish nature conservation, and in 2004 won the RSPB's Golden Eagle Award for the person who has done most for nature conservation in Scotland in the last 100 years. 'Getting on with it' is exactly what he has done. And it has brought extraordinary results.

As mentioned at the start of this book, the rewilding charity Trees for Life has also been busy bringing squirrels back to many of their former haunts. Becky Priestley began working with Roy in 2011 and learned a great deal about squirrel translocation during her time with him. She then went on to become the Red Squirrel Project Manager for Trees for Life. Using the same pioneering techniques and efficacious methods, she and the team have since relocated squirrels to a further nine Highland woodlands, from which they have been missing for decades.

This ambitious project has seen more than 170 squirrels released across seven sites in the Wester Ross area since 2016 – Ben Shieldaig, Coulin Estate, Plockton, Inverewe Gardens, Attadale, Letterewe and the Reraig peninsula – as well as at Spinningdale in Sutherland and, most recently, further south at the Ardtornish Estate at Lochaline on the Morvern peninsula.

As with Roy's Highland Wildlife Foundation project, only small numbers of squirrels are removed from any given site, leaving donor populations unaffected. Becky carries out the health checks in the field as the animals do not now have to go to a vet. This not only saves times but importantly minimises stress for the squirrels too.

The reintroductions have been a real success story, with all of the new populations breeding after release. They are now expanding throughout the available habitat and are also beginning to colonise additional woodlands some distance from the release sites. The charity's long-term aim is for many of the new populations to link up. Watching this expansion is an incredibly positive and satisfying aspect to Becky's work.

'It's fascinating to see how far squirrels will travel to colonise new woodlands. Although most of the squirrels remain local to where we release them, there are often one or two that venture much further afield – sometimes up to seventeen kilometres away, and over quite significant stretches of open ground. We are now beginning to see expansion on a much wider scale, and are delighted that additional populations have established in a number of locations – at Torridon from the Shieldaig release, Balmacara and Strome from the Plockton and Attadale releases, and Achnashellach from the Coulin release. There have also been occasional sightings around Gairloch, Lochcarron and Kinlochewe, so we can't wait to see how the populations continue to grow over the next few years.'

Recent surveys have shown that the Isle of Mull would be another suitable location to introduce reds in the future. If this becomes a possibility, it would give another important island population where there would be far less possibility of threats from encroaching greys.

The charity is keen for people to report sightings, to help with their monitoring of population expansion. 'Sightings are invaluable because they help us know whether squirrels are still present at the release sites, how far they are travelling to colonise new woodlands, and also whether they are breeding,' explains Becky. 'It's quick and easy to help the project in this way – simply visit the national sightings database at scottishsquirrels.org.uk and log your record.'

Taking part in citizen science to assist with monitoring is just one of many ways that people can contribute to squirrel projects all around the country, and in the case of Trees for Life, they are eager to involve local communities. Local residents often help to release the squirrels and feed them for a few months afterwards. This can be really beneficial as the animals acclimatise to their new surroundings. An education programme with local schools and guided walks in the release woodlands offer further opportunities for people to take part.

The difference that the work of Trees for Life and Roy Dennis are making cannot be underestimated – by significantly increasing both the numbers and the range of the red squirrel, their projects are a huge contribution to the conservation of this much-loved species. And with the continuation of such dedicated collaborative work and help from members of the public, landowners and conservation organisations and charities, the red squirrel has the prospect of a healthy future in the northwest.

23

Alladale

Alladale Wilderness Reserve in Sutherland is another place where important red squirrel translocation work has been taking place, but it's also a place where its owner and his work seem to instantly trigger a response and an opinionated conversation will develop.

When it comes to wildlife matters, as I have mentioned, there is often an extensive margin between viewpoints. The problem is that there may also be a dangerous void in the middle. Akin to a dark crevasse, it's easy to fall in and almost impossible to return. Perception is a curious thing. Arguments rage and instead of measured debate, there is heat and vitriol. This is likely to end in impasse whereby nothing is achieved. It's undoubtedly one of many reasons why it has become treacherous trying to find common ground between various groups and organisations. And as we have seen, there is a long list of animals and birds that cause instant verbal inflammation. The red squirrel has overcome this particular hurdle and remains one of the best-loved mammals in the British Isles, but we have a long way to go with many other species.

Even conservation groups and charities that should be working towards the same end goal have their particular agendas, and that frequently leads to disputes too. Given the biodiversity crisis we are facing, we should put aside all bigotry and instead pull together to save the natural world. Sadly, it's never that simple.

One of the finest examples of landscape-scale habitat restoration where numerous groups, individuals and organisations *are* working together is Cairngorms Connect. It is, according to their website,

> a successful partnership of neighbouring land managers, committed to a bold and ambitious 200-year vision to enhance habitats, species and ecological processes across a vast area within the Cairngorms National Park.

> The Cairngorms Connect area stretches over 600 square kilometres – it is a landscape of superlatives: ancient woodlands intersected by sparkling rivers and lochs, encircle an Arctic-like mountain massif – the most extensive and wildest of its kind in Britain; there are vast tracts of blanket bog, tranquil wetlands and secret woodland bogs. It is a place where eagles soar, wildcats prowl and red squirrels forage; home to plants, insects, birds and mammals found in few other places. The strength of Cairngorms Connect is the coming together of like-minded managers committed to delivering habitat enhancement at a scale unparalleled in Britain.

I love that idea and wish that we could follow this model in many other parts of the British Isles, for there is no doubt ours is one of the most beautiful countries in Europe, but it is changing at an alarming rate. And not for the better.

In the last twenty years or so, a growing number of extraordinary philanthropic individuals have been working tirelessly to fund landscape-scale habitat restoration projects in Scotland.

Danish couple Anders and Anne Holch Povlsen are Scotland's biggest landowners. Founders of Wildland, theirs is an overriding mission, a long-term vision of landscape-scale conservation over three large and significant Highland estates – one of which forms part of Cairngorms Connect. Their aim is to restore natural balances not only for wildlife but also for the benefit of small remote communities.

While the Povlsens are seldom in the press, and keen to avoid publicity, Paul Lister of Alladale conversely has received copious media coverage. He is often portrayed as a madcap philanthropist whose main mission is to return the wolf to Scotland. That is but a fractional part of a massive story – it's an infuriating trait of human nature that we hear what suits us then distort and modify the facts, which brings me back to perception. Frequently labelled the Mad Wolf Man, when I went to speak to him about his projects I discovered he couldn't care less what people say; in fact, I think he relishes it, so much so that he sometimes greets visitors while wearing a wolf mask. I have always appreciated people who have the ability to laugh at themselves.

Filled with anticipation and wondering if all I have heard about him is true, I am keen to make my own mind up and I am heading northwest to his remote 9,300-hectare Alladale Wilderness Reserve in Sutherland, on a drowsy afternoon at the tailend of summer. Moorland blushes with late heather, and thistledown fluff drifts on a breeze below wheeling red kites. You would be forgiven for thinking that the vast expanses of treeless moorland that sprawls across much of the route over the hills from Aberfeldy, up over the Dalnacardoch Pass and onto the main trunk road north, is the typical dramatic Scotland that tourists flock here to see. These bleak moorlands, riven with grouse-shooting butts and scarred with an ever-increasing network of hill tracks, are a travesty; they are for the most part devoid of life. Having walked on several of the big hills in this region that borders the fast-moving A9 between Perth and Inverness, I have seen the lack of life for myself.

I want to meet Paul Lister, because his work and his ethos intrigue me. I also want to see the environmental restoration work that is on going over a typical, remote, wet Scottish desert. How have the trees progressed during the period that Paul has owned the reserve? I imagine that it's a place where the habitat is changing for the better, and as mentioned in the last chapter, it's since been deemed a suitable environment in which to reinstate red squirrels. This in itself is positive news.

As I turn off at Ardgay, the road shrinks smaller and smaller. The route to Alladale follows the River Carron. Fat Cheviot ewes graze boggy fields, and stands of willow and alder fringe the riverbanks, dipping their roots in the fluctuating depths. Areas of commercial clear-fell forestry have left deep, ugly ruts on thin soils; stumps and hag lie like dead bodies, but foxgloves soothe the devastation. Gradually, ancient pines begin to dot the landscape as low rays of sun throw spot lighting onto spiders' webs. Soon there are more – gnarly sylvan behemoths. Most of them are at the very end of their long lives.

As I pass through a stone entrance and proceed on the long driveway up to Alladale Lodge, the pinewoods on either side of the road alter again – among the primitive forms youngsters burgeon forth – a soft green fuzz of new growth. Hope. I find cones chewed by red squirrels too, and large wood ants' nests. Below me on the cascading falls of the River Alladale sinuous salmon battle upstream to spawn, and Scotch Argus butterflies land on mauve heads of devil's bit scabious.

Alladale Lodge sits above the track commanding a vista of birch and pinewoods. I stand momentarily at the front door, absorbing a rapidly changing scene as low cloud folds in over the glen below. This austere building is now far removed from its initial raison d'être as a Victorian shooting lodge. Today visitors come to enjoy the tranquillity, to slow down and discover more about nature and habitat restoration, and how to contribute to saving the planet. That is, if it's not already too late. Alladale is a place where life-changing ideas, as well as dramatic ground-breaking conservation initiatives take flight.

Disappointingly, Paul Lister is not wearing a wolf mask as he greets me warmly. Brown and fit, he has the energy of a frenetic mustelid on steroids. He clearly looks after himself for, as he says without preamble, 'Time is running out for us, and I still have so much to do. We have to stand back and look at ourselves; we cannot keep doing the same things over and over again while expecting different results. We are at a crossroads in our evolution. We need to learn from this pandemic, or we are finished. We must now challenge the capital growth model and realise the mistakes we have made. The plant, fungal and non-human animal kingdom lives within a well-balanced trophic cascade but we humans are the complete opposite.'

For the next thirty-six hours, he embarks on a whistle-stop tour of both Alladale and his brilliant mind. His thoughts are black and white, and some of his ideas raise hackles. He flits through an astonishing range of projects, ideas and issues until I feel intox-icated, overwhelmed, but intrigued and yes, inspired. He's not merely someone with the gift of the gab; he puts his money where his mouth is and is making a difference as one of the world's most influential and active philanthropic conservationists and rewil-ding advocates. Not everyone would agree. Sometimes we don't like to be told that what we are doing is going to lead to disaster. Sometimes it's hard to face reality.

It's now seventeen years since Paul bought Alladale. 'Like most of the Highlands, it had suffered centuries of abuse, the great woodlands that once covered these glens had long gone and with

them all the life therein. Now in their place lies a degraded landscape of heather and bracken that we think is the norm. It's not
the bloody norm; it's a silent landscape devoid of life except deer
and sheep, and they cause untold damage.' He laughs loudly. 'My
vision is to restore that balance. We have planted nearly a million
trees here and will be planting more. We have dramatically reduced
deer numbers, allowing increased natural regeneration, but we
don't offer commercial stalking any longer. It's not our ethos to
refer to stalkers and to stalking so instead, I say rangers and culling
– vermin is another bloody awful out-dated word. What on Earth
does it mean anyway?' He laughs again. 'We shouldn't use it any
more. I see myself as a custodian of this landscape, and I want to
leave it in a better state. We have already restored vast areas of peat
land by blocking hill drainage. Healthy peat bogs as well as trees
are vital to help sequester carbon. We host groups of teenagers
through our HOWL project – Highland Outdoor Wilderness
Learning – because we believe that the future for conservation lies
in educating the younger generation. We have already returned
the red squirrel, and that has proved extremely successful. We
now have squirrels coming to the feeders in the garden; our guests
love to see them. We are also working closely with the Royal
Zoological Society of Scotland [RZSS] and other partners to
captive-breed the critically endangered wildcat. I would love to
see wildcats back here – it's the right place for them, it's where
they should be. There's so much to be done, and yes, time is
running out.'

He laughs a lot, despite the seriousness of his message. He tells
me that he is deeply, deeply ashamed of our past. Twenty years ago,
following the death of his father and mentor – the co-founder of
the UK's biggest and most successful furniture business, MFI – a
man he adored, he had a massive change of consciousness. It led
to a mission to leave a legacy to connect people and business to

nature and to increase the number of international environmental projects, in Romania, Italy, Spain and Belize.

'To give money is one thing, but to get involved and roll your sleeves up is so much more rewarding and effective. Have you any idea, Polly, that less than 3 per cent of charitable donations are used for the environment, climate change and wildlife? We are the ones that depend on the natural world, not the other way around. The UK is also the only country in the whole of Europe that doesn't have an apex predator. I want to bring wolves back in a controlled environment and learn from it. I do not advocate setting them free to roam at will – that is not what I want to do. I want to put them into a giant reserve surrounded by a fence where there would be deer too.'

And there hangs the problem – opposition to the fence that would make walking across the reserve difficult. Scotland has unique access laws. The Land Reform (Scotland) Act 2003 establishes statutory public rights of access to land, and many individuals, landowners, organisations and groups, including Ramblers Scotland and the John Muir Trust, have grave concerns about the erection of high electric fencing around an area as large as 20,000 hectares that would contravene the 2003 Act and impede public access. 'Suggestions that a zoo licence would be required, that would stipulate that predators and prey – wolves and deer – couldn't co-exist in the enclosure are totally wrong – we are talking about a vast wilderness area, not a zoo or safari park. There's simply no comparison – this is different and it would be wonderful. Surely members of the public and access groups would be excited about the possibility of seeing wolves and knowing that there was something bigger out there? There is a huge cultural barrier to their presence in this country. Farmers in particular hate them,' Paul says.

What is also of public concern is the possibility that vandals or activists might cut fences, as happened when thousands of farmed

mink were released during the 1960s and 1970s, with disastrous
results that we are still trying to rectify today, though a few wolves
are hardly comparable. The presence of a naturalised population
of mink, a voracious North American predator species, like the
grey squirrel, has had repercussions. Mink have caused terrible
damage to native fauna, in particular the little water vole, as well
as numerous ground-nesting birds. In winter, heavy falls of snow
around the fence area might also enable easy escape routes for
wolves. However, as Paul explained, any wolves would have elec-
tronic tags so they could be traced.

It is understandable that the public value Scotland's unique
access laws and are worried that if the project went ahead, it
would set a precedent for other similar ventures, or even exclusive
hunting zones where exotic animals were released to shoot, and
the public barred from entry.

People do not appear to be worrying that wolves would devour
hill walkers – something that has never happened anywhere. But
that too has of course been mooted. The impasse seems to be the
restricted access caused by a high fence. 'If the trial were allowed
to go ahead, we would soon witness the effect that wolves would
have on the crazily high deer population. Wolves would also
keep deer herds on the move so that natural regeneration could
take place, allowing the woodlands to recover. Wolves are now
in every country in Europe – we are the only ones without any
large carnivores. It has left us with an overcultivated dead zone.'
Not everyone agrees. It's a subject that is guaranteed to fuel strong
opinions either way. On a very personal note, I would be fearful
for the safety of the poor wolves, given humanity's horrendous
track record for illegal persecution.

Paul Lister's energy is such that at times I find it hard to keep
up. There are moments when I feel a need to lie flat somewhere
dark. Then we are off out again to look at another fantastic area

of the reserve. I am overwhelmed by what is happening. Albeit slowly due to the acidic ground, the weather and the rough terrain, the trees are rising again. Wild beauty engulfs me. I can feel the potential. It fuels me with hope and restores my flagging spirit as I listen to many things I find frightening. Our loss of biodiversity terrifies me.

Then I wander down alone to have a quiet peep at the wildcat project. Like the rest of the ventures at Alladale, it's impressive. Spacious enclosures naturalised with shrubs and small trees aid privacy. The cats are breeding successfully – proof that they are content in their incarceration.

Almost all the food at Alladale is home-produced or sourced locally. We have picked the fruit and vegetables earlier from Alladale's latest impressive venture – state-of-the-art aquaponic gardens, where filtered waste from organically fed trout fertilises the produce. We sit outside the lodge eating exotic salads with the latest arrivals, people from the south who have heard about the work at Alladale and in Europe and are eager to join forces in the growing movement to restore nature. Paul proudly tells them that the magical little twinflower, one of Scotland's rarest plants and a pinewood specialist, was recently found in the newly establishing woods. Then he tells them about the squirrels.

Each evening after dinner, Paul shows me wildlife films revealing some of his numerous projects. He is the founder of The European Nature Trust (TENT), whose work focuses on the restoration and protection of some of the world's finest natural sites. 'TENT also supports and co-produces these beautiful natural history and documentary films. Their story is a serious call to action.' Some are so beautiful that they move me to tears as they reveal the breadth of wildlife that still clings on in some of the world's last great natural havens. We have lost so much already, and yes, I agree with him. Time is running out.

'We have to move away from horrendous factory farming, move to a more plant-based diet – stop buying foods flown from the other side of the world. We cannot sustain the population growth – the UN suggests it's around 81 million people a year. We have to stop buying all this crap that we don't need. From the furniture business, I moved to conservation, so I speak from experience. It is the ultimate privilege to dedicate my resources to restoring natural habitats and fighting for environmental causes. Much of my time now is spent connecting people while persuading wealthy individuals and businesses to do the same. It's the only way forward. The top 1 per cent of the world's wealthiest families control 45 per cent of the world's capital – for heaven's sake, how sustainable can that be, where we now find ourselves making a mess of dealing with a pandemic? We had it coming to us.

'But look at that glen – even without trees, it's perfect wolf habitat. They would do so well on that ground. Think of the bare tundra; it's not so dissimilar is it? With all the crags and rocky areas in Alladale's glens, there are hundreds of suitable safe places for them to den too,' he says wistfully.

I too feel wistful as I drive away from Alladale. Staying here has made me think even more deeply about ways we can all work together to save nature – and ourselves. Golden afternoon light floods into the pinewoods and the ancient trees take on an auburn tinge. I stop and wander with my camera, noting the patchwork of colour on the boggy, peaty ground – a bog is never simply a bog. Coloured plant jewels and a treasury of mosses: jade, emerald, ruby, amber, and the bright yellow of bog asphodel, and thousands of tiny unseen organisms. No, a bog is never simply a bog. Skeletal grey forms of geriatric dying pines stand stretching tired limbs against a backdrop of stormy blue – like Native American totem poles, their patterns scribed by woodpeckers and invertebrates.

So what is my perception of Paul Lister? I wish the world had more maverick philanthropists like him. He is someone who achieves the impossible. And he knows that without nature we are nothing. We are not separate but an integral part of it. That is something that I repeat over and over again. We need to finally accept that basic fact. Is it really so hard to grasp?

I still haven't seen any squirrels, but during my pensive wander, I discover lots more squirrel-chomped cones – some consumed recently. The light begins to fade as the sun slips low in the sky, tawny grasses waltzing in the breeze whisper of autumn. I drive on down the glen. Two small boys on bikes wait in a layby for my car to pass. They wave cheerily. Dressed in shorts, they both have muddy knees. One of them has joined-up freckles and a big smile. I stop and reverse slowly back. 'Do you have red squirrels here? Do you ever see them?' I ask through the open window. Having first looked worried both boys begin to chatter animatedly, each interrupting the other as they tell me that yes, they often see squirrels. 'We were watching them today chasing one another round and round a tree in the garden. We love the squirrels because they are funny, and sometimes they make these wee chattery noises. Och, they're great wee beasties.' I thank them and drive away. In those last few words, I can hear the voice of an adult. 'Och, they're great wee beasties.' Squirrels – everyone loves them.

24

Foresting Hope for the Red Squirrel

Witnessing the astonishing transformations that have taken place at home on a small scale, I only wish that I had been able to do something dramatic on a really grand scale. My efforts seem so inadequate when I know how bad things have become. However, I also know if we are to beat climate change and the loss of our biodiversity, we have to remain positive. One particular wild place close to home with a healthy population of red squirrels has been luring me enticingly for the past two decades. It's a work in progress, abundant with life, where native Scots pines have been nurtured and now march forth up the hillsides, bringing hope for other such woodland ventures. It is not a large area, but already I can see it is making a difference.

It is soon after dawn on a milky-grey spring morning as I take to the hill. Here deer fencing has been removed, for it has played its part in enabling verdant natural regeneration to blossom on an ancient forest site. Now that the trees are fully established, the deer can return and will help to keep areas of the woodland open for other wildlife.

Shafts of watery sunlight massage distant snow-stippled hills, gradually painting an ocean of dark pines with fleeting gold filaments. In a clearing of frosted tawny grasses, a primal ritual is taking place. Blackcock are on their 'lek', displaying immaculate blue-black plumage enhanced by engorged crimson wattles as flashes of white under-feathers are revealed in a puff. They have been returning to this breeding ceremony arena for generations, to show off and compete with one another, backing and advancing in a series of choreographed manoeuvres. Their grey hens, high in a larch tree, feed on buds; plum pudding shapes against the grey pallor of daybreak. Like the black game, red squirrels are amorous too, chasing one another around a sun-tinted pine trunk temporarily glowing as orange-red as their pelts.

On the skyline a granny pine stands skeletal. Some of its great limbs lie collapsing at its base and a dark shape reveals a perching raven. Only its wispy throat hackles break the clean lines of its silhouette – a beard of feathers on the wisest member of the corvid family. It calls with a harsh *krrrk*. Its mate will already be on eggs for this is a bird, like the golden eagle and the crossbill, that usually nests early in the season. A curlew cries mournfully, adding grace notes to the distinctive bubbling sound of jousting blackcock. The wood is alive with amorous proclamation.

Crested tits have survived the long, hard winter and prospect a fissured pine bole. Dependent on myriad insects and seeds, they constantly search amongst the leaf litter, emitting their shrill *zee, zee, zee* calls. Parties of other tits and goldcrests may join them. Like the red squirrel, this rare member of the titmouse family stores food for the harsher months. Valuable invertebrate-rich dead wood bears the pockmarks of a great-spotted woodpecker's drilling. His rhythmic drumming echoes across the glen. Soon the cuckoo and willow warbler will arrive, adding new songs to the Scottish spring, and nesting golden plover may call plaintively from surrounding moorland. A buzzard mews. The blackcock will shortly retire to feed, continuing their breeding dance once more at sundown.

This small pinewood is exemplary. Here our three native species of conifer – Scots pine, juniper and yew – grow in profusion. Its open spaces are dappled with holly, rowan, ash, and bird cherry; the lower reaches by a burn fringed with contorted oak and hazel. There are dippers and tiny amber spotted trout. Willows and alders paddle in peaty pools where frogs come to spawn, and midges hatch forth in billions. We may find the ubiquitous Highland midge perilous with its power to drive us completely mad, but even this fiendish little beast is important for it provides a source of food for dozens of other insects, bats and birds.

As the days lengthen and the sun rises high in the sky, iridescent dragonflies will flit. Impressively large golden-ringed dragonflies dressed in their striped black and yellow favour the acidic burns and pools found on the woodland edge. These voracious predators feast on insects including midges, bees, wasps, damselflies and even other dragonflies. Fast flyers and on the wing from May through to early autumn, close examination reveals hypnotic emerald green eyes. Curious conical mounds conceal colonies of large wood ants, their laborious work symbiotic to the health of the woodland. When dew is heavy, the understorey of blaeberry, cowberry and heather is latticed with thousands of spiders' webs: lace in the dew, intricate works of art etched in silver by unseen grafters. In summer, foxgloves stand aloof, as the sun gently teases out nautilus shapes of unfurling ferns at their feet.

A sexton beetle, dapper in black and orange, stops atop a verdant cushion of moss. Bees and hoverflies will hum lazily as the rare pink-white twinflower shyly shows its faces, with wintergreens, wood anemone, wood sorrel, and orchids – twayblade and lady's tresses – as the season advances. Mixed flocks of chaffinches and siskins chatter in the canopy. A pair of Scottish crossbills is caught in a beam of light, the male's waistcoat flamboyantly glowing. These specialists, like the squirrels, are extraordinarily well-equipped to extricate the winged seeds from deep within tough pinecones, their crossed mandibles allowing them to scissor through to the heart of the matter.

Exposed roots are marked by pine marten scats; a badger has been excavating an anthill's sandy loam. The atmosphere is perfumed with earthiness, and the pungency of fox. Sometimes in winter I find otter tracks on snowy paths by the burn's edge, or their spraint on rocks fringed in spring with primrose, dog violet and celandine. During severe winter storms, mountain hares descend for shelter amid brown hares, and roe and red deer. One winter, an area of mature pines was used as a roost by a juvenile golden eagle and a sea eagle, their large elliptical pellets and white splats left as evidence on gale-scorched earth. Sometimes I saw the sea eagle drifting lazily over the woodland below, its vast wing-span making me catch my breath, its shadow engulfing all in its wake. There used to be wildcats here too – a very long time ago now – together with the great capercaillie, nicknamed 'horse of the woods'. As this secret wood flourishes and continues its miraculous cycle of constant death and renewal from season to season, they could return.

*

Twenty-one years on, my little wood at home is no longer in its infancy, and the wildlife is returning there too, but it's important to recognise that these places of nature are but islands in a sea of impoverished, poisoned land. Though our immediate population of red squirrels thrives, and birds including goldfinches and bullfinches nest in our thick hedgerows, and tree pipits and yellowhammers sometimes nest on the farm too, all these precious creatures are vulnerable. When I plant a tree there is satisfaction and the thought that I am doing something that will go on into the future – long, long after I am gone. It fills me with optimism that I am leaving something valuable behind. Though I vouch for the fact that if you plant, they will come, we need to plant more, stop draining our wetlands, restore peatlands and restore our traditional hay meadows and make it worthwhile for farmers to work with nature, and not against it. For now I have run out of space. I will have to put my efforts into supporting and promoting those who are working tirelessly to restore our beleaguered ecosystems elsewhere.

Planting trees is indeed one of the finest things we can do to reverse trends of rising greenhouse gas emissions from burning fossil fuels and global forest destruction. However, if we have no space, no land and nowhere to do this, we will likely feel powerless to contribute and think that it is someone else's job. This is not the case, for numerous inspirational organisations have landscape-scale planting schemes, and even planting vital flowers for pollinators in parks, small gardens or window boxes can contribute. Don't forget too that trees make beautiful gifts, gifts that keep growing for sometimes hundreds of years. And even a dead tree is of the utmost value. In the words of journalist Bryce Nelson, 'People who will not sustain trees will soon live in a world that cannot sustain people.'

We all love red squirrels and it's important to remember that without our help, we are in increasing danger of losing these incredible, delightful woodland sprites. And that thought is simply too painful to contemplate. For now we have to continue to work towards a safe future for the red squirrel in the forest – for in saving squirrels we will be saving so much more.

AFTERWORD

The Child in Nature

It's January, the season when a fox's hormones fuel the urge to breed. At night I lie in bed and hear their eerie yattering as they wander on their nocturnal forays, their soundtrack accompanied by tawny owls – they too are preparing for breeding. My son Freddy and I are following a lattice of worn tracks close to the farm, made as the deer trek up and down in the gloaming, to the richer pickings in the fields below. The wood is thick with their distinctive aroma. Now that the bracken has died away, and before the new growth, it's perfect for exploration. Badger paths are fascinating, and following them reveals another world. Their tracks are stippled with distinctive round snuffle holes marking the spots where they have grubbed for worms and beetles. There's been copious digging; examination of a new latrine pit reveals once again their catholic diet and the remains of all manner of invertebrates. A woodcock sprints into the air from the leaf litter, and we find pinfeathers, leftovers from a fox's feast – despite it binocular vision and cryptic plumage, another woodcock has been less lucky and has been caught unawares. Fieldfares leave their berry feast in a hurry amid exuberant clatter and chatter as they take to the sky in a flurry.

For as long as Freddy and I can remember, quiet forays into the natural world have been a precious part of our own. Even now that he is in his thirties and I am nearly sixty, they are as vital as during our childhoods – even more so. They provide peace and timeless absorption as we look, watch, smell and listen. There's no agenda; we go where the paths lead us.

We are the lucky ones, for we were both children of nature. Sometimes we spend long days together on the craggy, wind-scoured ridges of Scotland's mountains, walking all day, inhaling panoramas stretching far below. Eating our picnics in the lee of lichen covered rocks, listening to the rutting of the red deer during autumn, or the passing of whooper swans, their musical bugling calls emerging over the mountain summits, through ethereal inversions. Occasionally there is a fleeting flurry of ptarmigan, or we watch an eagle, or marvel at dragonflies living the high life in a bog pool. Often it is the tiny flowers beneath our boots, the perfume of crushed bog myrtle, or the vibrant colour of mosses and ferns that hold our attention. We relish days when squalls and storms reveal the wrath of the elements. It's enervating, inspiring and equally has the power to still. Even visits to a green space in the heart of a busy city can trigger that same glowing feeling that comes with a close encounter with the wild: a blackbird's melodic serenade, a freshly-emerged butterfly, wind in the hair, tadpoles, a red squirrel burying acorns in leaf litter, intricate patterns of lichen like maps on a crumbling gravestone, the brief glimpse of a vanishing weasel, spiders' webs twinkling with dew.

All tiny pieces of an unfathomable jigsaw.

Such adventures, either alone or in company, are integral to my ability to deal with life. It is a life that has become ever faster and more furious, in a mad world governed by screens and gadgetry, where we are like manic wasps bombarded with a ceaseless onslaught of information. It makes us flit crazily from subject to subject worrying, fretting under interminable pressure. Many of us crack under the strain. We have lost the ability to live in the moment and are always thinking of the next thing, and the next. And the next.

One in four adults and one in ten children will have a mentally-related illness during their lifetime. Mental health problems are universal and complex and manifest in dozens of different ways, from depression, angst and lack of self-esteem to eating disorders, drug and alcohol abuse, and self-harm. Some severe mental health issues are clinical, while others have a tangible cause. Many people exist in a state of flux, walking that fine line between coping, surviving and struggling. I was once one of those people.

The reason for my parents parting when I was a child was largely because my father was a chronic alcoholic, and though my mother protected me as best as she could, once they divorced, life changed. Being sent away to a boarding school that I loathed, and crippled with homesickness, I now endured the mayhem caused by Dad's drunkenness. Sometimes when I went to spend time with him during the school holidays, there were nights when he woke me in the small hours and we would blindly take to the road. His driving was terrifyingly erratic and afterwards, in a drunken stupor, he would cry like a child over the loss of my mother.

It led me to bouts of debilitating anxiousness and a form of depression. Freeing myself from the darkness seemed impossible. I wanted to make life better for Dad, and couldn't. During my teenage years, the situation worsened, because Dad's condition led to constant stress for us all. There was his erratic, loquacious behaviour and his long spells in rehab centres, numerous occasions when he didn't turn up or went missing for days, and no one seemed to know where he had vanished. I don't blame anyone – my parents both did the best they could under the circumstances. Perhaps it may come as a surprise when I reiterate that I feel so grateful to them that the incredible places where we lived, and the freedom that they gave me to wander, provided the most powerful tool to help deal with my unhappiness and instability: nature. Nature was reliable.

When I crossed the threshold of our door, I was swiftly absorbed into a far safer world. And Ardnamurchan, that great sea-girt rock jutting out west into the Atlantic Ocean where I was fortunate to spend my formative years, always provided me with solace. It became my rock when I found myself in hard places emotionally. It was, for me, a paradise with its stunted wind-sculpted oak woodlands, its shores rich with life, its clear waters that unveiled perfect marine gardens brimming with vibrant treasure, and the red deer, otters, eagles, ravens and seals, even wildcats. The hillsides and bogs, the steep dells with glades of oak stretching up the burnsides away from grazing pressures, were studded with wildflowers and filled with the songs of spring migrants and the calls of the cuckoo; they delighted, fascinated and grounded me. My mother was like me in this respect. On reflection, I wonder if perhaps she knew that letting me go so unfettered was the best way to deal with a situation over which she had little control. She knew how happy I was outside. Most importantly, my time in the wild taught me a basic fact: nature can salve myriad troubles, ease pain, yes, even physical pain, and soothe a tortured soul. It is absorbing and fascinating in turn.

> Our lives are over-crowded, over-excited, over-strained.
> We all want quiet. We all want beauty. We all need space.
> Unless we have it, we cannot reach that sense of quiet in
> which whispers of better things come to us gently.

When Octavia Hill, founder of the National Trust, wrote these words, she can have had no idea how they would resonate today.

Reading them now, 125 years later, they are more pertinent than ever.

Richard Louv in his seminal book, *Last Child in the Woods*, describes the phenomenon of Nature Deficit Disorder (NDD): the human cost of alienation from nature. Louv lists a diminished use of the senses, attention difficulties and higher rates of physical and emotional illnesses. He also states that our society is teaching people to avoid direct experience in nature. Research shows that spontaneous interaction with the natural world has declined by 90 per cent since the 1970s.

Dr William Bird, a GP and founder of Intelligent Health, a company he set up to promote physical activity, states: 'The outdoors can be seen as a great outpatient department whose therapeutic value is yet to be fully realised.' In an interview for *Outdoor Nation*, he said: 'Children who don't connect with nature before the age of twelve are less likely as adults to connect with nature. They therefore lose out on the resilience nature provides when you're really stressed.'

It's clear that nature's impact on our minds is dramatic and includes benefits to our cognitive functioning, which in turn leads to richer inter-personal relationships and enhanced behaviour traits. The fact that time spent in nature, regular exercise and fresh air also has such a beneficial impact on our physical fitness is without doubt. It has been proven to ease problems relating to allergies, while boosting the immune system and helping to lower blood pressure.

With a sharp escalation in people prescribed with antidepressants at astronomical cost to the National Health Service, prescribing nature as therapy for situational depression, for example, could prove far more effective.

So why have we become estranged from something we know provides society with countless benefits? Why do we eschew the services freely offered by the natural world? Dr William Bird says: 'People need to understand what they're missing out on – something really fundamental, a connection with the rest of life.'

The fact that time spent in nature, regular exercise and fresh air also have such a beneficial impact on our physical fitness is without doubt.

In 2012, the naturalist and author Stephen Moss produced a thorough report, *Natural Childhood*, for the National Trust. It is not only a thought-provoking, revelatory call to arms, it also explores positive ways forward to tackle the growing issues associated with NDD. It's not a nostalgic, romanticised suggestion of turning the clock back, but rather, it hopes to find ways that will lead to the nation's children safely playing outside once more. Moss states that although NDD is frequently viewed as a problem mainly relating to technology and poverty, and perhaps exacerbated in low-income urban areas, it applies to society as a whole. The report also explores the loss of safe play areas associated with a meteoric rise in traffic and the perceived threat of child abduction. Shockingly, he reveals that one of the most dangerous places for children to be is in their own homes. Here, risks are manifold: electricity, chemical poisoning, fire, knives, the internet, social media and many more. And worst of all perhaps is the fact that most incidences of child abuse are carried out by close relatives or family friends – at home. According to Richard Louv, children are banned continuously from various life-enhancing activities, leading to the criminalisation of natural play.

When I received that devastating telephone call one summer morning and my uncle told me that my father had taken his own life, it shattered my world. Fuelled by adrenaline, I seemed to sail through the next phase. It was some time before it all came back to haunt me. Suicide leaves a gaping wound. It leaves dozens of unanswered questions, the whys, the regrets, the wishing one had said and done so many things. And you cannot press rewind. The road ahead lay brooding and bleak. It was fraught with serious bouts of melancholia. Then, entirely due to my natural childhood and the time I spent with wildlife and my animals, I mustered an ability to come to terms with the premature loss of my father. I found a way to cope. Nature is my panacea, and its power is so reliable, so magical that we need to understand its crucial importance to each and every one of us. It is the reason why I reiterate over and over again that, 'without nature, we are nothing'.

In the words of Richard Louv, 'If we are going to save the environment, we must also save an endangered species: the child in nature.' This is where each of us has a vital role to play because we can all be the guiding hand leading the younger generation on a journey that will be our salvation.

So I urge you to venture out into a place where you know there are wild things great or small – even better if you are lucky enough to be able to find red squirrels or any of the other creatures that share their world. Sit quietly as you wait and watch, and you will discover the finest free therapy available on Earth. Remember too that nature is fragile and needs our utmost respect, that we hold the future of all our precious wild things in the palm of our hands. And remember: without nature, we are nothing.

Acknowledgements

Many people help both directly and indirectly during the course of writing a book, and many probably don't realise the extent to which they inspire and steer. It would be impossible to name everyone who has supported, encouraged and advised me during the writing of *A Scurry of Squirrels*, so the list that follows is necessarily incomplete.

Firstly, my son Freddy has always shown great enthusiasm and interest in all my wildlife ventures and is a constant support.

I want to thank retired Scottish SPCA Inspectors Martin Love, who first brought me wildlife casualties many years ago, and Don Wilson, who was the source of my first orphan squirrel kits. Special thanks for friendship and wise advice with rehabilitation work are due to Gay and Andy Christie of Hessilhead Wildlife Rescue, and to Colin Seddon, Romain Pizzi, Sheelagh McAllister April Sorley and John Fletcher. I am also very grateful to Romain for writing the Foreword to this book.

I am grateful to Paul Lister of Alladale Wilderness Reserve, Roy Dennis of the Roy Dennis Wildlife Foundation, Becky Priestley of Trees for Life, as well as Anne Hamilton, Debbie Shann and Tom Bowser, who gave valuable support through the writing phase.

Thanks to the team at my favourite book festival – Wigtown – including Adrian Turpin, Anne Barclay and Clare Nash. They have given me opportunities to talk about squirrels, pine martens and other wildlife during the festivals of past years.

I would also like to offer my sincere gratitude to my friends at the inspirational rewilding charity, SCOTLAND: The Big Picture. Director Pete Cairns commissioned me to write the words to accompany Neil McIntyre's world-class squirrel images in the book *The Red Squirrel – A Future in the Forest*. An extended version of the last chapter from that book forms part of my ending to this one, and I am grateful for permission to use material from it here. Thanks also to fellow directors and filmmakers Mat Larkin and Mark Hamblin for their time and patience in creating a short film depicting the importance of nurturing the wild. It's not every photographer who tolerates working with a sharp-clawed baby squirrel balancing on their head, but Mat manages it with ease and style!

Thanks as ever to my publisher, Hugh Andrew, and the hard-grafting team at Birlinn. Copy-editor Helen Bleck has been exceedingly helpful. Most importantly, I want to thank my editor, Andrew Simmons, whose excellent ideas and wonderful sense of humour mean it's always a joy to work with Birlinn.

To my partner Iomhair Fletcher, who brilliantly makes and modifies housing and enclosures for creatures various – usually at short notice – and helps with a constant round of feeding, cleaning and tending in our joint bid to nurture the wild: thank you from all the animals, and me.

And lastly, thanks to Cloudy, the red deer calf who lay beneath my makeshift desk keeping me company throughout the lockdown of summer 2020, peacefully chewing the cud while I wrote much of this book.

Polly Pullar
June 2021